László Ferencz
Daniela-Lucia Muntean

Virtual Screening and Docking Active Ingredients

László Ferencz
Daniela-Lucia Muntean

Virtual Screening and Docking Active Ingredients

Examples and Applications

LAP LAMBERT Academic Publishing

Impressum / Imprint

Bibliografische Information der Deutschen Nationalbibliothek: Die Deutsche Nationalbibliothek verzeichnet diese Publikation in der Deutschen Nationalbibliografie; detaillierte bibliografische Daten sind im Internet über http://dnb.d-nb.de abrufbar.

Alle in diesem Buch genannten Marken und Produktnamen unterliegen warenzeichen-, marken- oder patentrechtlichem Schutz bzw. sind Warenzeichen oder eingetragene Warenzeichen der jeweiligen Inhaber. Die Wiedergabe von Marken, Produktnamen, Gebrauchsnamen, Handelsnamen, Warenbezeichnungen u.s.w. in diesem Werk berechtigt auch ohne besondere Kennzeichnung nicht zu der Annahme, dass solche Namen im Sinne der Warenzeichen- und Markenschutzgesetzgebung als frei zu betrachten wären und daher von jedermann benutzt werden dürften.

Bibliographic information published by the Deutsche Nationalbibliothek: The Deutsche Nationalbibliothek lists this publication in the Deutsche Nationalbibliografie; detailed bibliographic data are available in the Internet at http://dnb.d-nb.de.

Any brand names and product names mentioned in this book are subject to trademark, brand or patent protection and are trademarks or registered trademarks of their respective holders. The use of brand names, product names, common names, trade names, product descriptions etc. even without a particular marking in this work is in no way to be construed to mean that such names may be regarded as unrestricted in respect of trademark and brand protection legislation and could thus be used by anyone.

Coverbild / Cover image: www.ingimage.com

Verlag / Publisher:
LAP LAMBERT Academic Publishing
ist ein Imprint der / is a trademark of
OmniScriptum GmbH & Co. KG
Heinrich-Böcking-Str. 6-8, 66121 Saarbrücken, Deutschland / Germany
Email: info@lap-publishing.com

Herstellung: siehe letzte Seite /
Printed at: see last page
ISBN: 978-3-659-78043-1

Contents

Foreword

New treatment options for patients means new drugs and biological products. National authorities for drugs approve each year innovative new products:

a) with *new chemical entities* never used before in medicine;

b) which contain previously approved or related substances in order to treat the same or different diseases.

The costs for new drugs are huge and under risk of failure even after approval for marketing. For this reason pharmaceutical companies often stop research on less promising drugs earlier and they need powerful research „instruments" to design correctly the study from the beginning. One of these instruments which is under continuous development is computer modelling of ligand – receptor interaction. A full understanding of drug – receptor interaction helps to make decisions which take into consideration the direct and indirect effect, desirable and undesirable effects, time of action etc. and, of course, it helps to build reliable modelling software.

The process of finding a new drug (or a new pesticide) against a chosen target for a particular disease (or against a certain pest), usually involves a screening procedure, wherein large libraries of chemicals are tested for their ability to modify a target, a protein or an enzyme. With the constant development of the drug screening technology, new screening methods and techniques have been developed.

An important function of these screenings is to show how selective the compounds are for the chosen target. The ideal is to find a molecule which will interfere with only the chosen target, but not with other, related targets. To this end, other screening runs will be made to see whether the "hits" against the chosen target will interfere with other related targets (cross-screening).

It is very unlikely that a perfect drug candidate will emerge from these early screening runs. It is more often observed that several compounds are found to have some degree of activity, and if these compounds share common chemical features, one or more pharmacophores can then be developed. At this point, medicinal chemists will attempt to use structure-activity relationships (SAR) to improve certain features of the lead compound.

A range of parameters can be used to assess the quality of a compound, or a series of compounds, as proposed in the Lipinski's Rule of Five. Such parameters include calculated properties such as XLogP3 to estimate lipophilicity, the number of H-bond donors, H-bond acceptors, molecular weight, polar surface area and measured properties, such as potency, in-vitro measurement of enzymatic clearance etc. Some descriptors such as ligand efficiency[1] (LE) and lipophilic efficiency[2,3] (LiPE) combine such parameters to assess drug likeness[4].

Computer-based methods are becoming increasingly important in studying the structure and function of biomolecules.

Molecular docking is a frequently used tool in structure-based rational drug design. Although early efforts were hindered by limited possibilities in computational resources, due to recent advances in high performance computing, virtual screening and docking methods became more and more efficient. These methods contributed to the development of several drugs and drug candidates that advanced to clinical trials[5].

This book was written to acquaint the reader with some of the possibilities of virtual screening for different ligands (drugs and pesticides) and docking them on the surface of different enzymes.

The examples were selected from these fields mainly due to the importance of finding new active ingredients, which can substitute old molecules with known adverse effects. When choosing and formulating the questions, their genuineness was sought out, most of them being adapted from papers published by the authors in

---

[1] A.L. Hopkins, C.R. Groom, A. Alex A., *Drug Discov. Today*, **2004**, *9* (10): 430–1.

[2] T. Ryckmans, M.P. Edwards, V.A Horne, A.M Correia, D.R. Owen, L.R. Thompson, I. Tran, M.F. Tutt, T. Young, *Bioorg. Med. Chem. Lett.*, **2009**, *19* (15): 4406–9.

[3] P.D. Leeson, B. Springthorpe, *Nat. Rev. Drug Discov.*, **2007**, *6* (11): 881–90.

[4] https://en.wikipedia.org/wiki/Drug_discovery, retrieved 2 August **2015**.

[5] A. Rudnitskaya, B. Török, M. Török, *Biochem Mol Biol Educ*, **2010**, *38* (4): 261-265

periodicals[6,7,8,9]. Even though the computation results do not perfectly fit the experimental values or their most recent interpretations, the examples were selected so as to illustrate a way of approaching the experimental behavior on the basis of theoretical treatment and within the limits of certain considered approximations.

Tg.-Mureş, on 25. August 2015   *The Authors*

---

[6] L. Ferencz, D. L. Muntean, *Farmacia*, **2015**, *63* (3): 422-428.

[7] L. Ferencz, D. L. Muntean, *Farmacia*, **2015**, *63* (2): 189-195.

[8] L. Ferencz, D. L. Muntean, *Rev. Roum. Chim.*, **2014**, *59* (9): 733-738.

[9] L. Ferencz, D. L. Muntean, *Acta Universitatis Sapientiae, Agriculture and Environment*, **2015**, in press.

1. The method

We used relatively simple tools in order to demonstrate that these procedures are easy to fulfill. The tests were run on a dual core Celeron© PC, using docking programs that are distributed free for personal use.

Hardware: Asus X401A PC, CPU Dual Core Intel© 820, 1.7GHz, 4 GB RAM.

Software: AutoDock Tools 1.5.6 Molecular Graphics Laboratory The Scripps Research Institute[10], AutoDock Vina by Sargis Dallakyan, The Scripps Research Institute[11], Open Babel 2.3.2.[12], PyRx 0.8, PubChem Compound Database. The operation system was the 64 bit version of Windows 7. We also used the Chem 3D Ultra 10.0, ChemDraw Pro 10.0 programs (under personal license) and the browser Mozilla Firefox.

The Similar Compounds search type of the Chemical Structure Search of the PubChem Compound Database[13] allows to locate records that are similar to a chemical structure query, using pre-specified similarity thresholds. Similarity is measured using the Tanimoto equation and the PubChem dictionary-based binary fingerprint. This fingerprint consists of series of chemical substructure "keys". Each key denotes the presence or absence of a particular substructure in a molecule. The fingerprint does not consider variation in stereo chemical or isotopic information. Collectively, these binary keys provide a "fingerprint" of a particular chemical structure valence-bond form.

The degree of similarity is dictated by the Threshold parameter. A threshold of "100%" effectively acts as an "exact match" to the provided chemical structure query (ignoring stereo or isotopic information), while a threshold of "0%" would return all

---

[10] G.M. Morris, R. Huey, W. Lindstrom, M.F. Sanner, R.K. Belew, D.S. Goodsell and A.J. Olson, *J. Comput. Chem,* **2009**, *16*: 2785-2791.

[11] O. Trott, A.J. Olson, *J. Comput. Chem.*, **2010**, *31*:455-461.

[12] N.M. O'Boyle, M. Banck, C.A. James, C. Morley, T. Vandermeersch, G.R. Hutchison, *J. Cheminf.*, **2011**, *3*:33-47.

[13] National Center for Biotechnology Information. PubChem Compound Database; http://pubchem.ncbi.nlm.nih.gov/search/search.cgi#

chemical structures in the PubChem Compound database. Various predefined thresholds between 99-60% are allowed.

Searching the databases (with over 30 million entries) is possible for a broad range of properties, including chemical structure, name fragments, chemical formula, molecular weight, XLogP, and hydrogen bond donor and acceptor count. PubChem can be accessed for free through a web user interface:

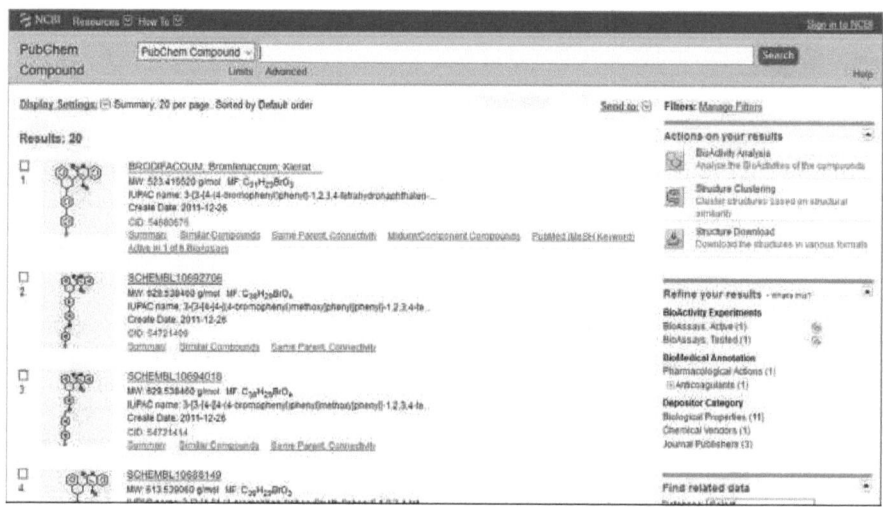

Fig. 1. The PubChem Compound Database's graphical user interface (GUI)

For docking, we used PyRx[14] and AutoDock Vina, which significantly improves the average accuracy of the binding mode predictions compared to AutoDock 4.

For its input and output, Vina uses the same PDBQT molecular structure file format used by AutoDock. PDBQT files can be generated (interactively or in batch mode) and viewed using MGLTools. Other files, such as the AutoDock and AutoGrid parameter files (GPF, DPF) and grid map files are not needed.

When using a flexible docking engine, then minimizing the input conformation of the ligands can reduce problems that are known to occur in conformer generation inside the docking engine, that arise if the input 3D conformation is not relaxed into good bond lengths and angles. For small molecules a good choice is to use some

---

[14] S. Dallakyan, A.J. Olson, *Methods Mol. Biol.* **2015**, *1263*:243-50.

molecular mechanics to optimize the structure down to a local energy minima, like UFF or mm2.

The assignment of Universal Force Field (UFF) atom types and the calculation of the molecular connectivity (identifying bonds, angular, torsional and inversion terms) has been performed using the routines available in the Open Babel package[11,12]. OpenBabel can be used for refining initial geometries with UFF molecular-mechanics optimizations, adding or removing hydrogens to PDB protein files, and it has many other utility tasks that often arise in molecular modeling projects:

- support for a huge variety of common chemical file formats, including SDF/MOL, Sybyl, mol2, PDB, SMILES, XYZ, CML,
- batch conversion for multiple molecules in one file (e.g., splitting, merging, batch operation),
- automatic feature perception (rings, bonds, hybridization, aromaticity),
- complete programmer's toolkit including C++, Perl, Python interfaces for easy custom software development.

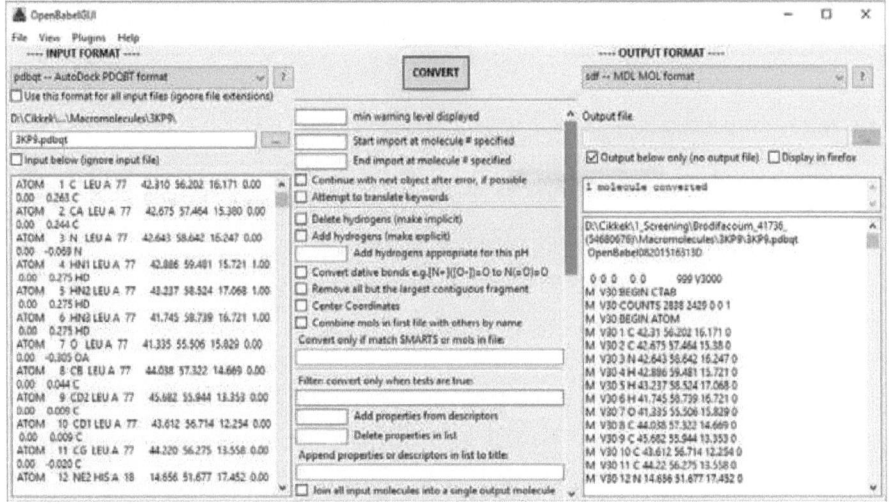

Fig. 2. The OpenBabel GUI

Open Babel supports a number of force fields which can be used for energy evaluation as well as energy minimization. We used the following energy minimization parameters: Conjugate Gradients optimization algorithm, 200 total number of steps, stop if energy difference is less than 0.1 kcal/mol.

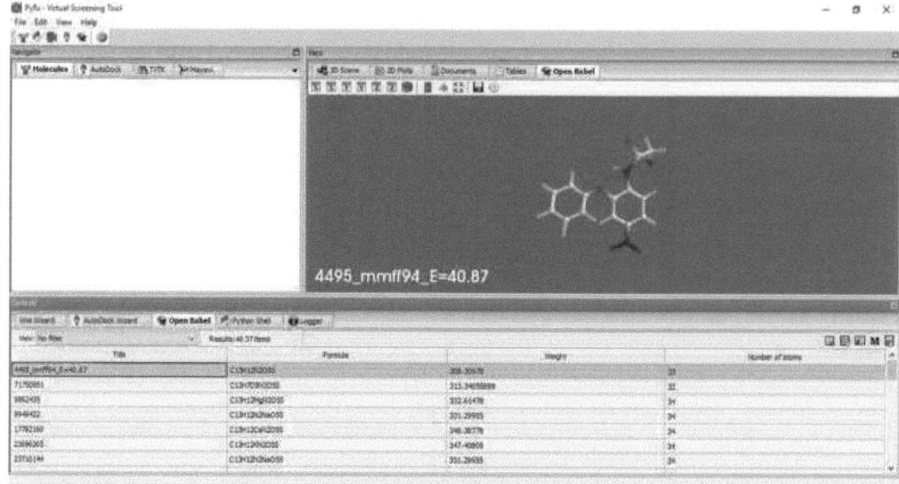

Fig. 3. The OpenBabel interface on PyRx

The docking calculation consists of a number of independent runs, starting from random conformations. Each of these runs consists of a number of sequential steps. Each step involves a random perturbation of the conformation followed by a local optimization (using the Broyden-Fletcher-Goldfarb-Shanno algorithm) and a selection in which the step is either accepted or not.

We used the default docking parameters:

– number of binding modes: 9,

– exhaustiveness (thoroughness of search): 8.

Larger values increase the probability of finding the global minimum, but also extend the computational time. Increasing the exhaustiveness value increases the time linearly and decreases the probability of not finding the minimum exponentially. Apart from exhaustiveness influenced by users, Vina has an internal heuristic algorithm to extend the search in accordance with an increasing number of atoms and rotatable bonds[15].

---

[15] L.K. Wolf, *Chem. Eng. News*, **2009**, 87:31.

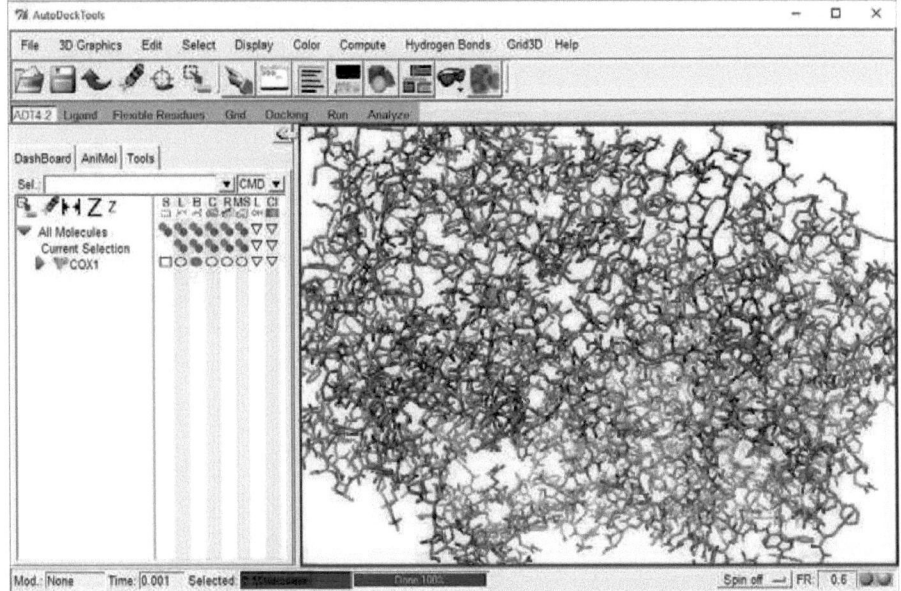

Fig. 4. MGL Tools 1.5.6 and AutoDock Tools

Each local optimization involves many evaluations of the scoring function as well as its derivatives in the position-orientation-torsions coordinates. The number of evaluations in a local optimization is guided by convergence and other criteria. The number of steps in a run is determined heuristically, depending on the size and flexibility of the ligand and the flexible side chains. However, the number of runs is set by the exhaustiveness parameter. Since the individual runs are executed in parallel, where appropriate, exhaustiveness also limits the parallelism. Unlike in AutoDock 4, in AutoDock Vina each run can produce several results: promising intermediate results are remembered. These are merged, refined, clustered and sorted automatically to produce the final result [16,17,18,19,20]. Vina creates *_out.pdbqt files where it stores all docked poses and scores[11].

---

[16] M.F. Sanner, *J. Mol. Graphics Mod.*, **1999**, *17*:57-61.

[17] M.F. Sanner, J.-C. Spehner, A.J. Olson, *Biopolymers*, **1996**, *38* (3):305-320.

[18] C. Bajaj, S. Park S, A. Thane: "A Parallel Multi-PC Volume Rendering System", ICES and CS Technical Report, University of Texas, **2002**.
http://www.cs.utexas.edu/~bajaj/cvc/software/docsTRLIB/PMVR.pdf

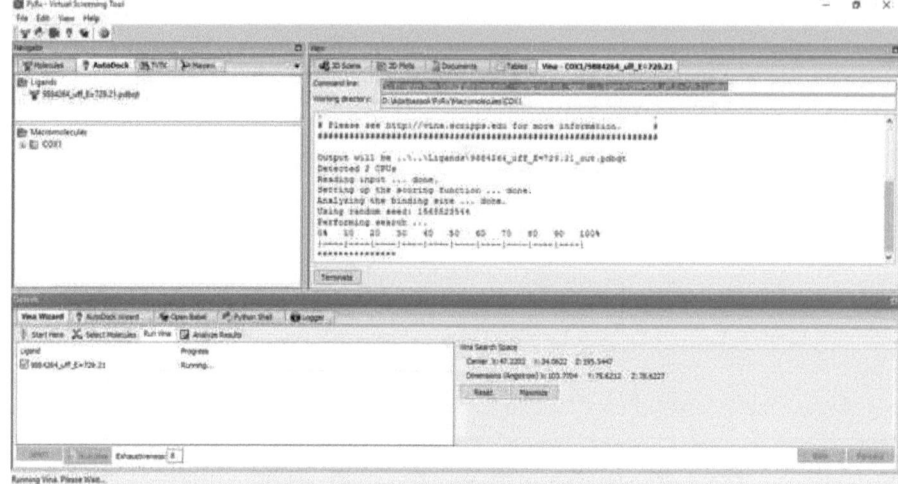

Fig. 5. PyRx and Vina Wizard. Performing search.

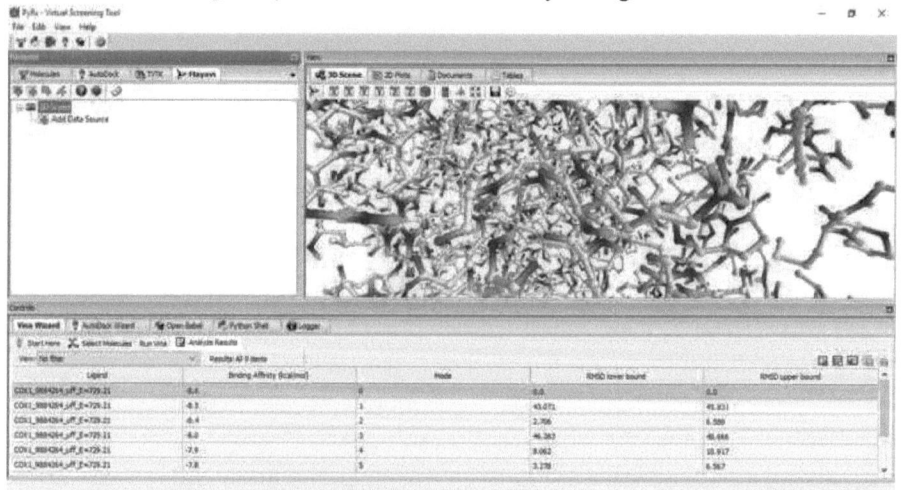

Fig. 6. Vina Wizard: analyzing the results

---

[19] C. Bajaj, V. Pascucci, D. Schikore: "Fast IsoContouring for Improved Interactivity", Proceedings of ACM Siggraph/IEEE Symposium on Volume Visualization, ACM Press, San Francisco, CA, **1996**, p. 39-46.

[20] M.F. Sanner, D. Stoffler, A.J. Olson: "ViPEr a Visual Programming Environment for Python". 10th International Python Conference, Virginia, **2002**.
http://www.scripps.edu/sanner/html/papers/IPC02.pdf

The predicted binding affinity of bound structures is given in kcal/mol. To compare the accuracy of the predictions of the experimental structure, AutoDock Vina uses a measure of distance between the experimental and predicted structures, RMSD, root-mean-square deviation.

RMSD values are calculated relative to the best mode and using only movable heavy atoms. For scoring, AutoDock Vina uses a united-atom function, which involves only the heavy atoms.

Two variants of RMSD metrics are provided by the software, rmsd/lb (RMSD lower bound) and rmsd/ub (RMSD upper bound), differing in how the atoms are matched in the distance calculation:

– *rmsd/ub* matches each atom in one conformation with itself in the other conformation, ignoring any symmetry;

– *rmsd'* matches each atom in one conformation with the closest atom of the same element type in the other conformation (rmsd' can not be used directly, because it is not symmetric);

– *rmsd/lb* is defined as follows: $rmsd/lb(c_1, c_2) = max((rmsd'(c_1, c_2), rmsd'(c_2, c_1))$.

For example, a highly symmetrical rigid ligand could be rotated relative to a reference conformation such that the new conformation is exactly equivalent to the reference conformation (think of a benzene ring flipped 180 degrees).

However, the internal numbering of atoms does not change during the docking run, so the rmsd/ub algorithm, which matches identically labeled atoms (rather than similar or equivalent atoms) would yield a significant rmsd, whereas the rmsd/lb algorithm would yield a more realistic rmsd of zero in this case.

For very flexible, asymmetric molecules, rmsd/lb would likely give unreasonably low rmsd's and rmsd/ub would be a better model.

2. PABA antagonists

Sulfanilamide (4-amino-benzenesulfonamide) was synthesized in 1908 and researchers proceeded to synthesize over 4500 sulfonamides by 1948. They are still an attractive group of drugs since they are very cheap. When they were first discovered, their mechanism of action was unknown. Observations that they structurally resemble *para*-amino benzoic acid (PABA, 4-aminobenzoic acid) prompted some scientists to suggest a connection. But it was not until the role of PABA in living cells was fully elucidated that the mechanism of action of sulfonamides was fully understood.

Sulfonamides are structural analogs and competitive antagonists of *para*-aminobenzoic acid and thus prevent normal bacterial utilization of PABA for the synthesis of folic acid. More specifically, sulfonamides are competitive inhibitors of dihydropteroate synthase[21] (DHPS, EC:2.5.1.15), the bacterial enzyme responsible for the incorporation of PABA into dihydropteroic acid, the immediate precursor of folic acid. Folic acid is important as one carbon source in many essential biochemical pathways. Sensitive microorganisms are those that must synthesize their own folic acid; bacteria that can utilize preformed folate are not affected. Bacteriostasis induced by sulfonamides is counteracted by PABA competitively. Sulfonamides do not affect mammalian cells by this mechanism, since they require preformed folic acid and cannot synthesize it. They are, therefore, comparable to sulfonamide-insensitive bacteria that utilize preformed folate.

Sulfonamides are bacteriostatic and they have a broad spectrum, but show poor activity against pseudomonas, enterococci and anaerobes. They are slow to act, since several generations are needed before appreciable depletion of folate pool and inhibition of growth. They are considered antimetabolites. Allergies to sulfonamide are common[22], hence medications containing sulfonamides are prescribed carefully.

---

[21] M.E. Cuff, J. Holowicki, R. Jedrzejczak, T.C. Terwilliger, E.J. Rubin, K. Guinn, D. Baker, T.R. Ioerger, J.C. Sacchettini, A. Joachimiak, Journal: to be published.
http://www.rcsb.org/pdb/explore/explore.do?structureId=4HB7 . Retrieved 6 January **2014**.
[22] Sulfa Drug Allergy, http://allergies.about.com/od/medicationallergies/a/sulfa.htm . Retrieved 2 May **2014**.

Because of this, it may be interesting to find similar compounds in order to enlarge the spectrum of DHPS inhibitors used today in therapeutics. We used the Similar Compounds search type of the Chemical Structure Search of the PubChem Compound Database, to locate records that are similar to the chemical structure of PABA, using pre-specified similarity thresholds. Using the threshold $\geq$ than 95% for the similar structures criteria, we found 15 compounds that meet these criteria.

We calculated the binding affinities for the ligands (including for p-aminobenzoic acid) to the surface of dihydropteroate synthase. Fifteen substances (with PubChem Compound ID = 22143252, 69285308, 2762775, 21063639, 22143251, 14332036, 302680, 45084823, 21446355, 11159424, 54746218, 95888, 69030033, 4465519, 154880) are better ligands for DHPS than PABA.

These ligands are shown in *Table 1* and below, together with PABA (CID 978).

Table 1. The 15 ligands with a similarity threshold $\geq$ 95%, and PABA (CID 978)

PubChem CID	IUPAC Name	Molecular Formula	MW [g/mol]
22143252	4-(sulfinoamino)benzoic acid	$C_7H_7NO_4S$	201.19978
69285308	4-amino-2,3-diiodobenzoic acid	$C_7H_5I_2NO_2$	388.92904
2762775	4-amino-2,3-difluorobenzoic acid	$C_7H_5F_2NO_2$	173.116906
21063639	4-(hydroperoxyamino)benzoic acid	$C_7H_7NO_4$	169.13478
22143251	1-carboxy-4-(sulfinatoamino)benzene	$C_7H_6NO_4S^-$	200.19184
14332036	4-amino-2-iodobenzoic acid	$C_7H_6INO_2$	263.03251
302680	4-amino-2-fluorobenzoic acid	$C_7H_6FNO_2$	155.126443
45084823	4-(fluoroamino)benzoic acid	$C_7H_6FNO_2$	155.126443
21446355	4-(chloroamino)benzoic acid	$C_7H_6ClNO_2$	171.58104
11159424	4-aminobenzenecarboperoxoic acid	$C_7H_7NO_3$	153.13538
54746218	4-(oxidoamino)benzoic acid	$C_7H_6NO_3^-$	152.12744
95888	4-(hydroxyamino)benzoic acid	$C_7H_7NO_3$	153.13538
69030033	4-(phosphanylamino)benzoic acid	$C_7H_8NO_2P$	169.117682

4465519	(4-carboxyphenyl)azanium	$C_7H_8NO_2^+$	138.14392
154880	4-carboxybenzenediazonium	$C_7H_5N_2O_2^+$	149.1268
978	4-aminobenzoic acid	$C_7H_7NO_2$	137.13598

Except for CID 22143252 (with 6 H-bond donors, instead of 5), all substances accomplish the Lipinski Rule of Five[23], also known as the Pfizer's rule of five or simply the Rule of five (RO5), which is a rule of thumb to evaluate drug likeness or determine if a chemical compound with a certain pharmacological or biological activity has properties that would make it a likely orally active drug in humans:

1. no more than five hydrogen bond donors.

2. no more than ten hydrogen bond acceptors.

3. a molecular mass under 500 daltons.

4. an octanol-water partition coefficient, LogP value under five.

Candidate drugs that conform to the RO5 tend to have lower attrition rates during clinical trials and hence they have an increased chance of reaching the market.

Table 2. The 15 ligands with a similarity threshold ≥ 95%, and PABA (CID 978)

PubChem CID	H-bond Donor	H-Bond Acceptor	XLogP3-AA
22143252	6	3	1
69285308	2	3	1
2762775	2	5	1
21063639	3	5	1
22143251	2	6	0.4
14332036	2	3	1.4
302680	2	4	1.2
45084823	2	4	2

[23] C.A. Lipinski, F. Lombardo, B.W. Dominy, P.J. Feeney, *Adv. Drug Deliv. Rev.,* **2001,** *46* (1-3): 3–26.

21446355	2	3	2.3
11159424	2	4	0.9
54746218	2	4	1.6
95888	3	4	1.7
69030033	2	3	2
4465519	2	2	0.8
154880	3	1	2.8
978	2	3	0.8

22143252

69285308

2762775

21063639

22143251

14332036

302680

45084823

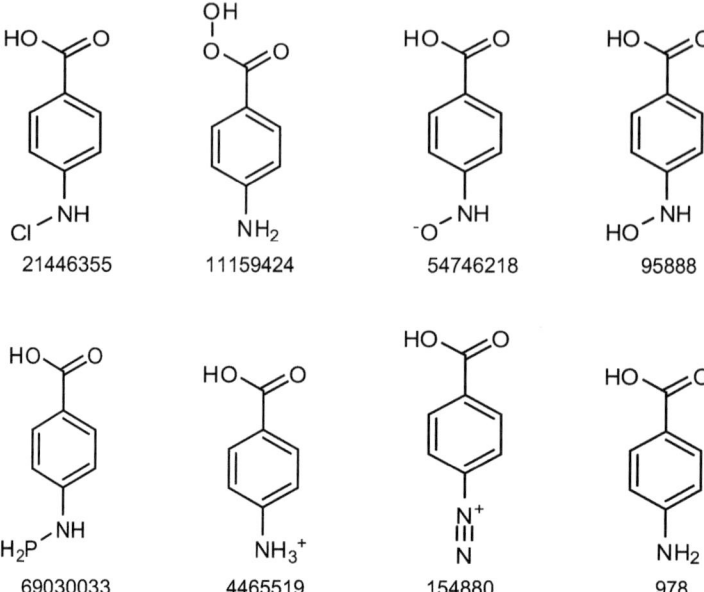

Table 3 presents the calculated binding affinities in descending order for the ligands and the enzyme DHPS (with PDB code 4HB7[21]).

Table 3. The calculated binding affinities greater than for PABA, in descending order for the enzyme dihydropteroate synthase (4HB7).

Enzyme-Ligand	Binding Affinity [kcal/mol]	rmsd/ub [Å]	rmsd/lb [Å]
4HB7_22143252_uff_E=105.40	-5.7	0	0
4HB7_69285308_uff_E=115.18	-5.6	0	0
4HB7_2762775_uff_E=71.95	-5.6	0	0
4HB7_21063639_uff_E=94.73	-5.6	0	0
4HB7_22143251_uff_E=101.25	-5.5	0	0
4HB7_14332036_uff_E=85.98	-5.5	0	0

4HB7_302680_uff_E=69.37	-5.5	0	0
4HB7_45084823_uff_E=66.69	-5.5	0	0
4HB7_21446355_uff_E=72.58	-5.5	0	0
4HB7_21063639_uff_E=94.73	-5.4	5.361	1.86
4HB7_11159424_uff_E=70.37	-5.4	0	0
4HB7_54746218_uff_E=68.72	-5.4	0	0
4HB7_95888_uff_E=81.41	-5.4	1.463	0.047
4HB7_95888_uff_E=81.41	-5.4	0	0
4HB7_21063639_uff_E=94.73	-5.3	29.271	28.19
4HB7_69030033_uff_E=-90.91	-5.3	0	0
4HB7_54746218_uff_E=68.72	-5.3	4.711	2.141
4HB7_4465519_uff_E=256.10	-5.3	0	0
4HB7_154880_uff_E=251.98	-5.2	26.716	25.983
4HB7_154880_uff_E=251.98	-5.2	0	0
4HB7_22143252_uff_E=105.40	-5.2	30.549	29.962
4HB7_22143252_uff_E=105.40	-5.2	4.151	3.367
4HB7_22143251_uff_E=101.25	-5.2	10.546	9.762
4HB7_54746218_uff_E=68.72	-5.2	4.933	2.162
4HB7_95888_uff_E=81.41	-5.2	4.705	2.176
4HB7_21446355_uff_E=72.58	-5.2	26.593	25.956
4HB7_978_uff_E=63.75	-5.2	0	0

The virtual screening results show that 15 compounds (with 0 rmsd values in *Table 3*) are strong inhibitors of DHPS (4HB7). The structure of enzyme was retrieved with Chem 3D's Online Find Structure from PDB ID option, and transformed to PDBQT with AutoDock.

The RMSD cutoff of 2Å is usually used as criteria of the correct bound structure prediction[24]. Using this cutoff value, the two metrics used for RMSD (summarized in *Table 3*) indicate that 15 compounds are better ligands of DHPS than PABA (CID 978), because they require less energy for binding. This suggests, that these substances will successfully substitute PABA and will act similarly with sulfonamides.

We used the default docking parameters:

– number of binding modes: 9,

– exhaustiveness (thoroughness of search): 8.

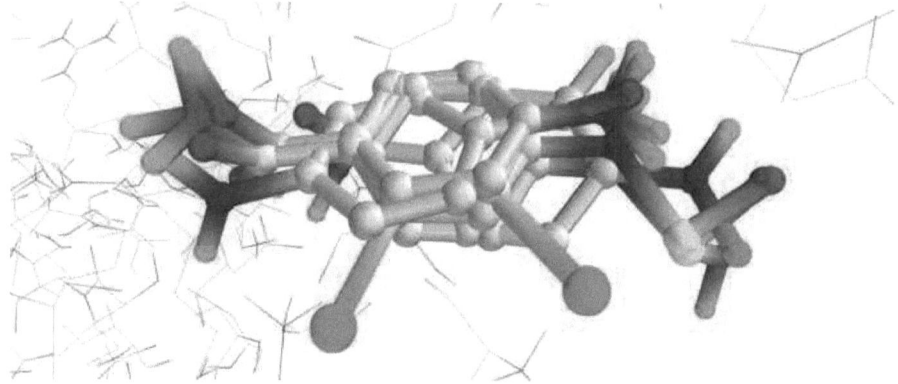

Fig. 7. Molecular docking of PABA and other 6 other ligands (CID 22143251, 22143252, 4876, 9971, 16461, 69285308) in protein target (Mayavi).

With reference to (solid) diazonium salts (like CID 154880), especially diazonium chlorides, they are often dangerously explosive, but diazonium salts with weakly coordinating anions are quite stable. For example, tetrafluoroborates can be stored almost indefinitely at room temperature and decompose gently when heated.

---

[24] B.D. Bursulaya, M. Totrov, R. Abagyan, C.L. Brooks, *J. Comput. Aid. Mol. Des.*, **2003**, *17* (11):755-763.

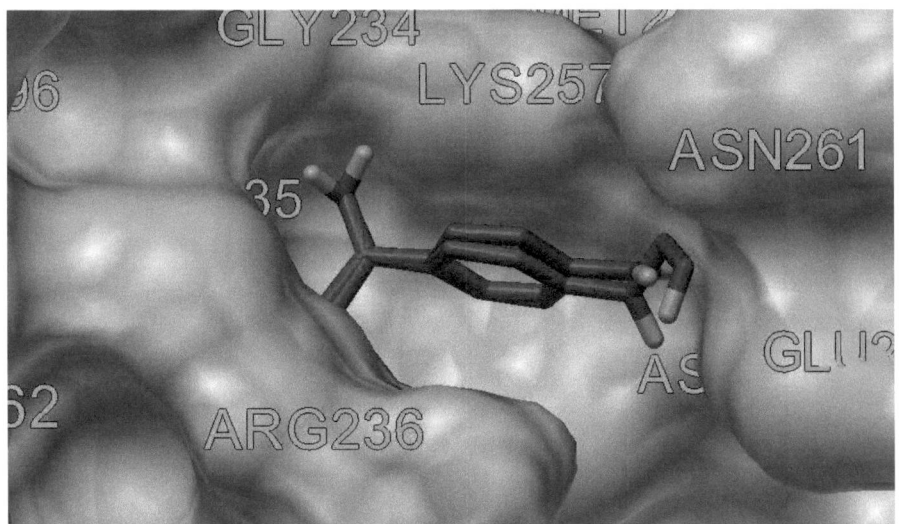

Fig. 8. Ligand CID 21063639 and PABA on the binding site of DHPS (Autodock).

In conclusion, compounds with PubChem ID: 22143252, 69285308, 2762775, 21063639, 22143251, 14332036, 302680, 45084823, 21446355, 11159424, 54746218, 95888, 69030033, 4465519 and 154880 have better binding affinity to dihydropteroate synthase enzyme than PABA, they present the correct bound structure prediction, so they seem to act alike, to be good substitutes for sulfonamides, and probable lacking their adverse effects. Further investigations are needed to establish their pharmacodynamic properties and toxicity.

3. Acetylsalicylic acid and similar compounds

Acetylsalicylic acid (2-acetyloxybenzoic acid, PubCHem Compound ID 2244, CAS 50-78-2) is an analgesic, antipyretic, antirheumatic, and anti-inflammatory agent. In countries where Aspirin is a registered trademark owned by Bayer, the generic term is acetylsalicylic acid (ASA).

2244

Acetylsalicylic acid mode of action as an anti-inflammatory and antirheumatic agent may be due to inhibition of synthesis and release of prostaglandins. Acetylsalicylic acid appears to produce analgesia by virtue of both a peripheral and central nervous system effect. Peripherally, acetylsalicylic acid acts by inhibiting the synthesis and release of prostaglandins. Acting centrally, it would appear to produce analgesia at a hypothalamic site in the brain, although the mode of action is not known. Acetylsalicylic acid also acts on the hypothalamus to produce antipyresis; heat dissipation is increased as a result of vasodilation and increased peripheral blood flow. Acetylsalicylic acid antipyretic activity may also be related to inhibition of synthesis and release of prostaglandins.

The analgesic, antipyretic, and anti-inflammatory effects of acetylsalicylic acid are due to actions by both the acetyl and the salicylate portions of the intact molecule as well as by the active salicylate metabolite. Acetylsalicylic acid directly and irreversibly inhibits the activity of both types of cyclooxygenase (COX-1 and COX-2) to decrease the formation of precursors of prostaglandins and thromboxanes from arachidonic acid. This makes acetylsalicylic acid different from other non-steroidal anti-inflammatory drugs (NSAIDS), such as diclofenac and ibuprofen, which are reversible inhibitors.

Acetylsalicylic acid is used in the temporary relief of various forms of pain, inflammation associated with various conditions (including rheumatoid arthritis, juvenile rheumatoid arthritis, systemic lupus erythematosus, osteoarthritis, and ankylosing spondylitis), and is also used to reduce the risk of death and/or nonfatal myocardial infarction in patients with a previous infarction or unstable angina pectoris[25].

Aspirin should not be taken by people who are allergic to ibuprofen or naproxen, or who have salicylate intolerance or a more generalized drug intolerance to NSAIDs, and caution should be exercised in those with asthma or NSAID-precipitated bronchospasm. Aspirin use has been shown to increase the risk of gastrointestinal bleeding. Patients with hemophilia or other bleeding tendencies should not take aspirin or other salicylates. Aspirin is known to cause hemolytic anemia in people who have the genetic disease glucose-6-phosphate dehydrogenase deficiency, particularly in large doses and depending on the severity of the disease. People with kidney disease, hyperuricemia, or gout should not take aspirin because it inhibits the kidney ability to excrete uric acid, and thus may exacerbate these conditions[26,27]. Aspirin should not be given to children or adolescents to control cold or influenza symptoms, as this has been linked with Reye's syndrome[28].

The popularity of aspirin declined after the market releases of paracetamol (acetaminophen, N-(4-hydroxyphenyl)acetamide) in 1956 and ibuprofen (brufen, 2-(4-(2-methylpropyl)phenyl)propanoic acid) in 1969. In the 1960s and 1970s, John Vane and others discovered the basic mechanism of aspirin effects[29], while clinical trials and other studies from the 1960s to the 1980s established aspirin efficacy as an anticlotting agent that reduces the risk of clotting diseases. Aspirin sales revived considerably in the last decades of the 20th century, and remain strong in the 21st century, because of its widespread use as a preventive treatment for heart attacks and strokes.

However, other similar compounds must be found to replace or to complete the action of this active substance, or to enlarge the spectrum of prostaglandin-endoperoxide synthase inhibitors used today in therapeutics.

---

[25] http://www.drugbank.ca/drugs/DB00945. Accessed 11 May 2014.

[26] M. Raithel, H.W. Baenkler, A. Naegel, F. Buchwald, H.W. Schultis, B Backhaus, S. Kimpel, H. Koch, K. Mach, *J. Physiol. Pharmacol.*, 2005, *56* (5), 89–102.

[27] H.T. Sørensen, L. Mellemkjaer, W.J. Blot, G.L. Nielsen, F.H. Steffensen, J.K McLaughlin, J.H. Olsen, *Am. J. Gastroenterol.*, 2005, *95* (9): 2218–24.

[28] S. Macdonald, *Brit. Med. J.*, 2002, *325* (7371): 988.

[29] J. Vane, *Nat. New Biol.*, 1971, *231(25)*: 232–235.

Using the Chemical Structure Search of the PubChem Compound Database and a threshold ≥ than 99% for the similar structure criteria, we detected 14 compounds that meet these criteria. We calculated the binding affinities for the ligands (including acetylsalicylic acid) to the surface of COX-1 and COX-2 enzymes. The structures of enzymes were retrieved from Protein Data Bank[30,31,32] in PDB format with Chem 3D's "Online Find Structure from PDB id" option. The water molecules, other small molecules, like solvent molecules and other relics of the isolation and crystallization procedures were removed.

X-ray crystallography usually does not locate hydrogens, hence most PDB files do not include them. But hydrogens, particularly those that can form hydrogen bonds, are important in binding ligands, so hydrogens were added to backbone N, and to amine and hydroxyl side chains. Atoms were renumbered, and PDBQT files generated with AutoDock Tools 1.5.6.

14 substances (with PubChem Compound ID = 298996, 94717, 22619484, 21481530, 95938, 94716, 3055063, 301846, 54224926, 301958, 19049630, 198203, 135269, 10176491) are better ligands for COX-1 and COX-2 than acetylsalicylic acid. These ligands are shown in *Table 4* and below, together with acetylsalicylic acid (CID 2244). All substances accomplish the Lipinski Rule of Five[23].

All 14 substances have 1 H-bond donor and 4 H-bond acceptors, and LogP values under 3.1:

Table 4. The 14 ligands with a similarity threshold ≥ 99%, and acetylsalicylic acid (CID 2244)

PubChem CID	IUPAC Name	Molecular Formula	MW [g/mol]	XLogP3-AA
298996	2-(4-methylpent-3-enoyloxy)benzoic acid	$C_{13}H_{14}O_4$	234.24786	2.8
94717	2-hexanoyloxybenzoic acid	$C_{13}H_{16}O_4$	236.26374	3.1

[30] http://www.rcsb.org

[31] D. Picot, P.J. Loll, R.M. Garavito, *Nature*, **1994**, *367*: 243-249.

[32] R.G. Kurumbail, A.M. Stevens, J.K. Gierse, J.J. McDonald, R.A. Stegeman, I.Y. Pak, D. Gildehaus, J.M. Miyashiro, T.D. Penning, K. Seibert, P.C. Isakson, W.C Stallings, *Nature*, **1996**, *384*: 644-648.

22619484	2-[(E)-but-2-enoyl]oxybenzoic acid	$C_{11}H_{10}O_4$	206.1947	2
21481530	2-(2,2-dimethylpropanoyloxy)benzoic acid	$C_{12}H_{14}O_4$	222.23716	2.6
95938	2-propanoyloxybenzoic acid	$C_{10}H_{10}O_4$	194.184	1.7
94716	2-butanoyloxybenzoic acid	$C_{11}H_{12}O_4$	208.21058	1.7
3055063	2-heptanoyloxybenzoic acid	$C_{14}H_{18}O_4$	250.29032	3.6
301846	2-(5-methylhex-4-enoyloxy)benzoic acid	$C_{14}H_{16}O_4$	248.27444	3.1
54224926	2-(2-methylpropanoyloxy)benzoic acid	$C_{11}H_{12}O_4$	208.21058	2.2
301958	2-(3-methylbut-2-enoyloxy)benzoic acid	$C_{12}H_{12}O_4$	220.22128	2.6
19049630	2-but-3-enoyloxybenzoic acid	$C_{11}H_{10}O_4$	206.1947	1.9
198203	2-(2-methylprop-2-enoyloxy)benzoic acid	$C_{11}H_{10}O_4$	206.1947	2.2
135269	2-pentanoyloxybenzoic acid	$C_{12}H_{14}O_4$	222.23716	2.6
10176491	2-prop-2-enoyloxybenzoic acid	$C_{10}H_8O_4$	192.16812	1.8
2244	2-acetyloxybenzoic acid	$C_9H_8O_4$	180.15742	1.2

298996

94717

23

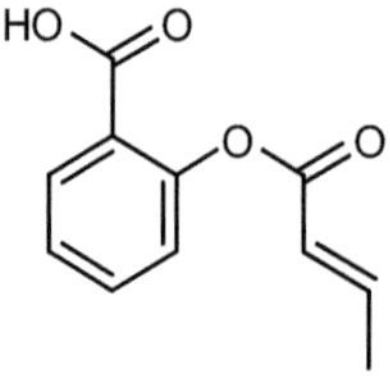

22619484

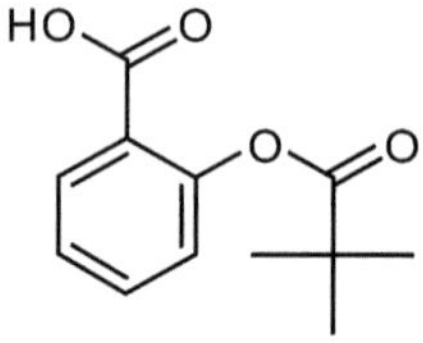

21481530

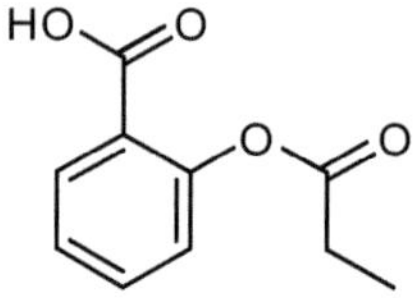

95938

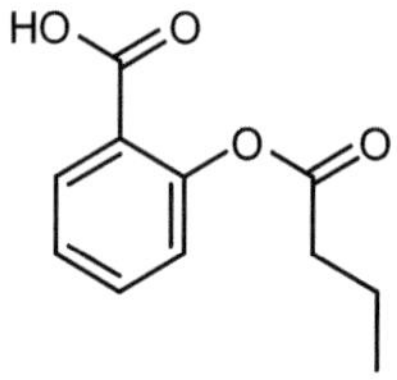

94716

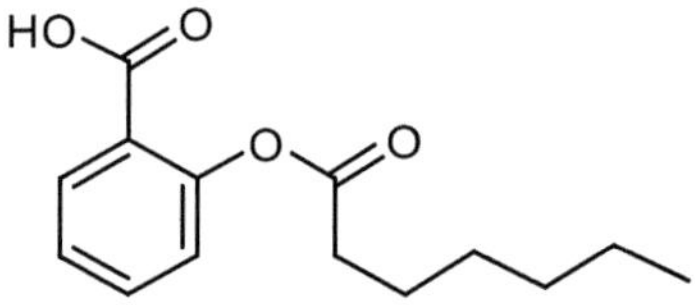

3055063

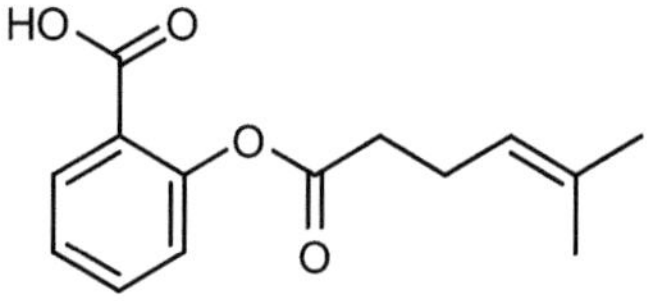

301846

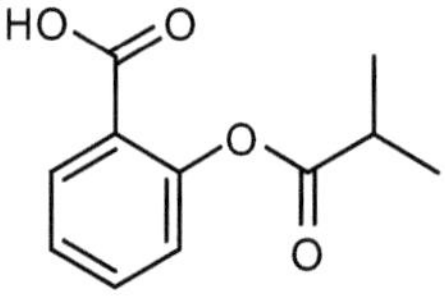

54224926

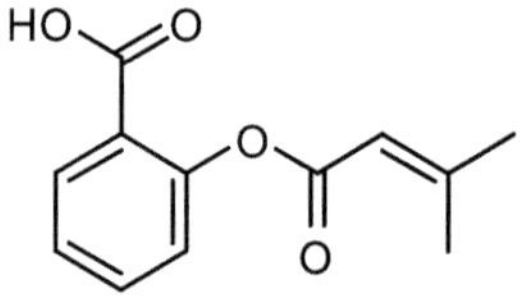

301958

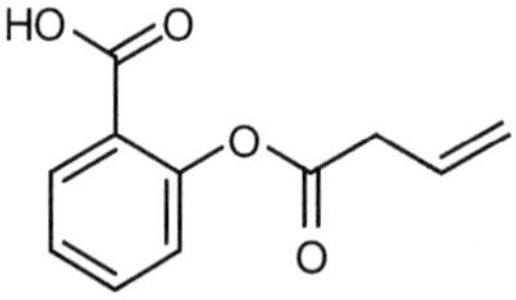

19049630

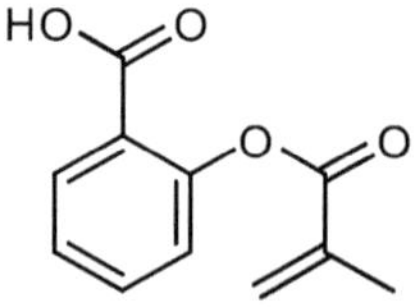

198203

24

135269

10176491

Table 5 and *Table 6* present the calculated binding affinities in descending order for the ligands and the two enzymes (COX-1 and COX-2), only for rmsd/ub and rmsd/lb = 0.

We used the following energy minimization parameters: Conjugate Gradients optimization algorithm, 200 total number of steps, stop if energy difference is less than 0.1 kcal/mol. The virtual screening results show that the 14 compounds are strong inhibitors of COX-1, and COX-2 (*Table 5* and *6*).

Table 5. The calculated binding affinities greater than for acetylsalicylic acid, in descending order for Cyclooxygenase-1 (COX-1).

Enzyme-Ligand (ligand's energies with Babel, kcal/mol)	Binding Affinity [kcal/mol]	rmsd/ub [Å]	rmsd/lb [Å]
COX1_301846_uff_E=195.59	-7.3	0	0
COX1_301958_uff_E=196.55	-7	0	0
COX1_298996_uff_E=155.58	-7	0	0
COX1_22619484_uff_E=436.89	-6.8	0	0
COX1_21481530_uff_E=152.88	-6.8	0	0
COX1_94717_uff_E=142.53	-6.8	0	0
COX1_95938_uff_E=132.22	-6.7	0	0
COX1_94716_uff_E=137.52	-6.7	0	0
COX1_3055063_uff_E=215.11	-6.6	0	0

COX1_54224926_uff_E=141.55	-6.5	0	0
COX1_19049630_uff_E=477.49	-6.4	0	0
COX1_198203_uff_E=138.43	-6.4	0	0
COX1_135269_uff_E=158.16	-6.4	0	0
COX1_10176491_uff_E=128.76	-6.2	0	0
COX1_2244_uff_E=419.53	-6.1	0	0

Table 6. The calculated binding affinities greater than for acetylsalicylic acid, in descending order for Cyclooxygenase-2 (COX-2).

Enzyme-Ligand (ligand's energies with Babel, kcal/mol)	Binding Affinity [kcal/mol]	rmsd/ub [Å]	rmsd/lb [Å]
COX2_298996_uff_E=155.58	-8	0	0
COX2_94717_uff_E=142.53	-7.5	0	0
COX2_301958_uff_E=196.55	-7.5	0	0
COX2_198203_uff_E=138.43	-7.3	0	0
COX2_301846_uff_E=195.59	-7.3	0	0
COX2_22619484_uff_E=436.89	-7.3	0	0
COX2_54224926_uff_E=141.55	-7.2	0	0
COX2_21481530_uff_E=152.88	-6.9	0	0
COX2_95938_uff_E=132.22	-6.8	0	0
COX2_10176491_uff_E=128.76	-6.8	0	0

COX2_3055063_uff_E=215.11	-6.6	0	0
COX2_94716_uff_E=137.52	-6.4	0	0
COX2_135269_uff_E=158.16	-6.3	0	0
COX2_19049630_uff_E=477.49	-6.2	0	0
COX2_2244_uff_E=419.53	-6.1	0	0

We used the default docking parameters: 9 number of binding modes, and exhaustiveness (thoroughness of search): 8.

Using the 2Å cutoff value, the two metrics used for RMSD indicate that all predictions for tested compounds are very accurate. Therefore, results indicate that 14 compounds are much better ligands of COX-1 and COX-2 than acetylsalicylic acid (CID 2244), because they require less energy for binding. This suggests that these substances will successfully substitute acetylsalicylic acid:

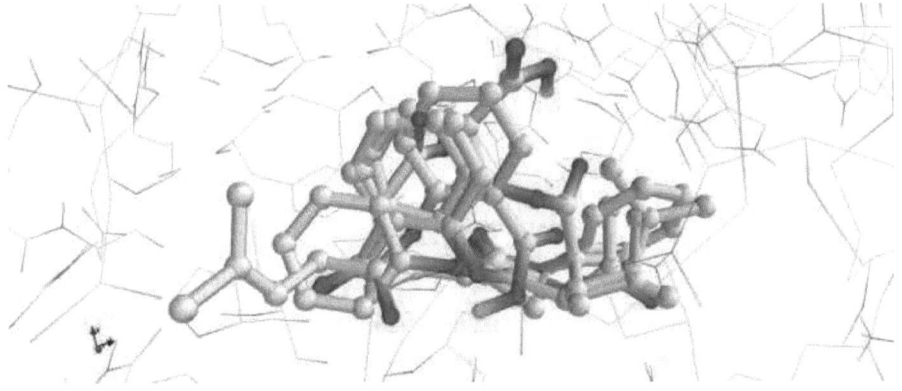

Fig. 9. Molecular docking of acetylsalicylic acid and other 6 ligands (CID 135269, 298996, 301846, 301958, 21481530, 22619484) in protein target (COX-1) for rmsd/ub and rmsd/lb = 0.

For example, all calculated binding modes (scenarios) of CID 298996 have greater binding affinities than acetylsalicylic acid (*Table 7*):

Table 7. Binding affinities and rmsd values for 298996 on the surface of COX-2.

Enzyme-Ligand	Binding Affinity [kcal/mol]	rmsd/ub [Å]	rmsd/lb [Å]
COX2_298996_uff_E=155.58	-8	0	0
COX2_298996_uff_E=155.58	-8	6.133	3.11
COX2_298996_uff_E=155.58	-7.7	44.88	43.181
COX2_298996_uff_E=155.58	-7.6	6.215	4.197
COX2_298996_uff_E=155.58	-7.4	45.977	43.85
COX2_298996_uff_E=155.58	-7.4	2.993	1.305
COX2_298996_uff_E=155.58	-7.2	46.297	43.652
COX2_298996_uff_E=155.58	-7.1	30.924	29.535
COX2_298996_uff_E=155.58	-7	32.545	31.01

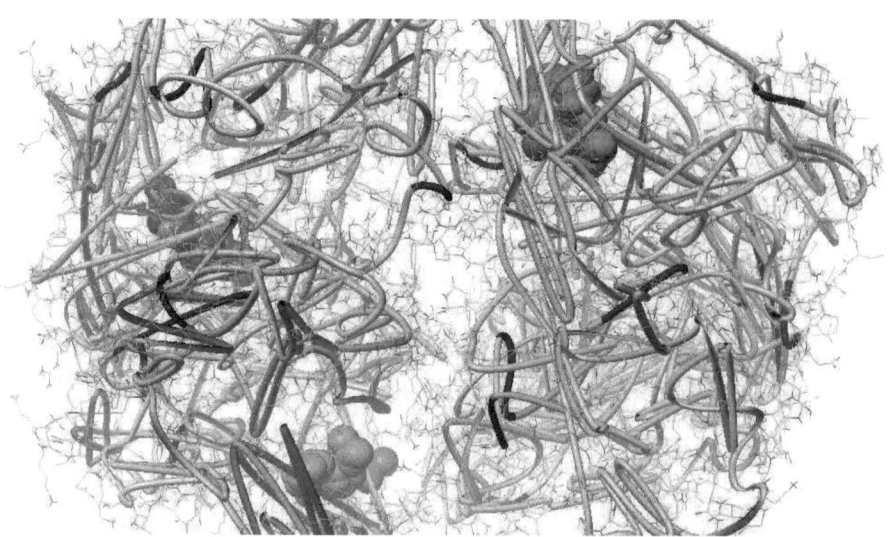

Fig. 10. Binding scenarios of CID 298996 on the surface of COX-2. Protein is represented with lines and beaded ribbons.

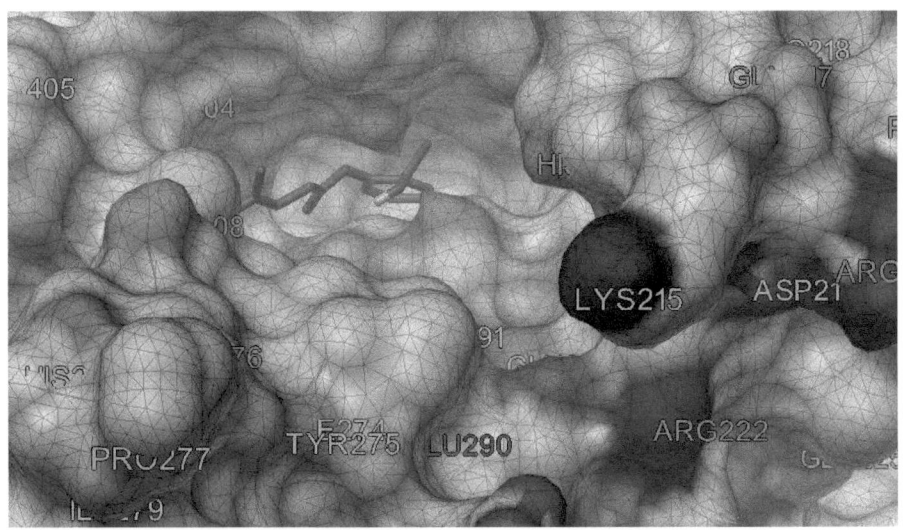

Fig. 11. Ligand CID 298996 (sticks) on the binding site of COX-2. Enzyme surface represented with Gouraud-shaded polygons is colored by polarity.

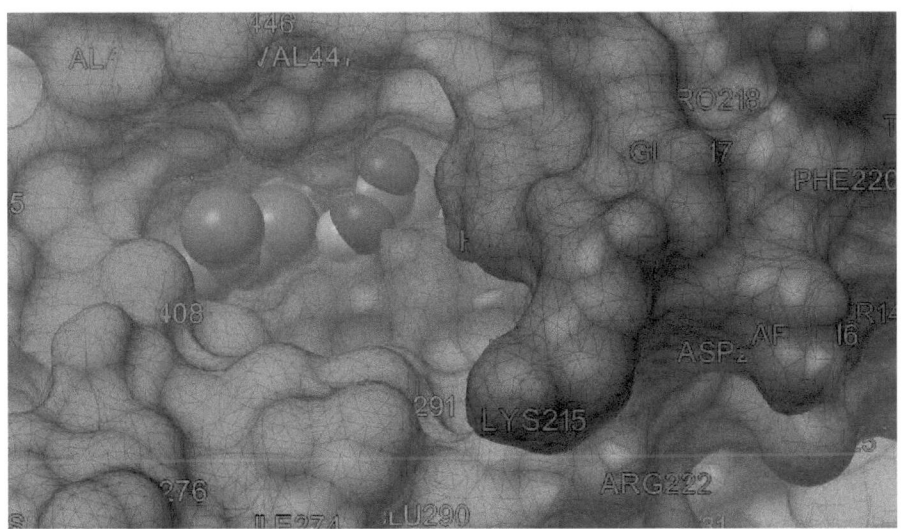

Fig. 12. Ligand CID 298996 (balls) on the binding site of COX-2, rainbow representation (AutoDock)

So we can state that compounds with PubChem ID: 298996, 94717, 22619484, 21481530, 95938, 94716, 3055063, 301846, 54224926, 301958, 19049630, 198203, 135269 and 10176491 have better binding affinity to cyclooxygenase enzymes (COX-1 and COX-2) than acetylsalicylic acid, they present the correct bound structure prediction, so they seem to be good substitutes for acetylsalicylic acid. Further investigations are needed to establish their pharmacodynamic properties and toxicity.

4. Nimesulide, a COX-2 selective inhibitor

Nimesulide is a non-steroidal antiinflamatory drug (NSAID), reported to be a selective inhibitor of cyclooxygenase-2 (COX-2), with analgesic and antipyretic properties.

Nimesulide is a relatively cyclooxygenase-2 (COX-2) selective, non-steroidal anti-inflammatory drug (NSAID) with analgesic and antipyretic properties. Its approved indications are the treatment of acute pain, the symptomatic treatment of osteoarthritis and primary dysmenorrhea, but its use in EU is limited due to reports of hepatotoxicity. Its mechanism of action is multifactorial and is characterized by a fast onset of action, giving a unique and broad action on inflammatory processes. Nimesulide was the first marketed NSAID which inhibits selectively COX-2, and belongs to a class of compounds (sulfonamides), that is unique among commercially available NSAIDs.

Nimesulide is different from other selective COX-2 inhibitors and classical non-steroidal anti-inflammatory drugs (NSAIDs). The anti-inflammatory effect mechanism of nimesulide (inhibition of inflammatory mediators) is similar to other classic NSAIDs, but Nimesulide exhibits a superior gastrointestinal safety as compared to other NSAIDS. It is known that nimesulide prevents NSAID-induced ulcers, while celecoxib and rofecoxib, which are more selective to COX-2, failed to prevent these ulcers. Nimesulide produces gastro-protective effects via a completely different mechanism. In addition, while selective COX-2 inhibitors increase the risk for cardiovascular diseases, nimesulide does not exert significant cardiotoxicity. These data suggest that gastrointestinal side effects of classic NSAIDs cannot be related to the COX-1 inhibition alone and also suggest that nimesulide is an atypical NSAID, which is different from both non-selective and selective COX-2 inhibitors[33,34].

The European Medicines Agency has completed a review of the safety and effectiveness of systemic medicines containing nimesulide (capsules, tablets, suppositories and powder or granules for oral suspension). The Agency's Committee

---

[33] H. Suleyman, E. Cadirci, A. Albayrak, Z. Halici, *Curr. Med. Chem.* **2008**, *15* (3):278-83.

[34] K.D. Rainsford K.D, *Curr. Med. Res. Opin.* **2006**, *22* (6): 1161–70.

for Medicinal Products for Human Use (CHMP) concluded that the benefits of nimesulide used systemically continue to outweigh its risks but that its use should be restricted to the treatment of acute pain and primary dysmenorrhea. It issued a recommendation that it should no longer be used for the treatment of painful osteoarthritis[35].

However, other similar compounds must be found to replace this active substance, or to enlarge the spectrum of prostaglandin-endoperoxide synthase inhibitors used today in therapeutics.

Using the Chemical Structure Search of the PubChem Compound Database and a threshold ≥ than 95% for the similar structures criteria, we detected 14 compounds that meet these criteria. Substances are shown in *Table 8*, together with nimesulide (CID 4495).

Table 8. The identified substances with a similarity threshold ≥ 95% and nimesulide (CID 4495)

PubChem CID	IUPAC Name	Molecular Formula	MW [g/mol]
10020363	N-[2-(2,4-difluorophenoxy)-4-nitrophenyl]methanesulfonamide	$C_{13}H_{10}F_2N_2O_5S$	344.29070
10902415	N-(4-nitroso-2-phenoxyphenyl)methanesulfonamide	$C_{13}H_{12}N_2O_4S$	292.31038
11493409	N-(2-hexoxy-4-nitrophenyl)methanesulfonamide	$C_{13}H_{20}N_2O_5S$	316.3733
11680642	N-(4-nitro-2-propoxyphenyl)methanesulfonamide	$C_{10}H_{14}N_2O_5S$	274.29356
11822507	N-[4-(hydroxyamino)-2-phenoxyphenyl]methanesulfonamide	$C_{13}H_{14}N_2O_4S$	294.32626
40489928	methylsulfonyl-(4-nitro-2-phenoxyphenyl)azanide	$C_{13}H_{11}N_2O_5S^-$	307.30184
4495	N-(4-nitro-2-phenoxyphenyl)methanesulfonamide	$C_{13}H_{12}N_2O_5S$	308.30978
4553	N-(2-cyclohexyloxy-4-nitrophenyl)methanesulfonamide	$C_{13}H_{18}N_2O_5S$	314.35742
58701679	N-(4-nitro-2-phenoxyphenyl)ethanesulfonamide	$C_{14}H_{14}N_2O_5S$	322.33636

[35] European Medicines Agency: Questions and answers on the outcome of the review of nimesulide-containing medicines, London, 16 October **2009**. Doc. Ref. EMEA/263700/2008

69899876	N-(3-cyclohexyloxy-4-nitrophenyl)methanesulfonamide	$C_{13}H_{18}N_2O_5S$	314.35742
71163623	4-nitro-2-phenoxy-1-(sulfinatoamino)benzene	$C_{12}H_9N_2O_5S^-$	293.27526
71364034	1,1,1-trifluoro-N-(4-nitro-2-phenoxyphenyl)methanesulfonamide	$C_{13}H_9F_3N_2O_5S$	362.28117
71365841	1-fluoro-N-(4-nitro-2-phenoxyphenyl)methanesulfonamide	$C_{13}H_{11}FN_2O_5S$	326.300243
71771973	N-methylsulfonyl-N-(4-nitro-2-phenoxyphenyl)methanesulfonamide	$C_{14}H_{14}N_2O_7S_2$	386.40016
9884264	N-[2-(4-methoxyphenoxy)-4-nitrophenyl]methanesulfonamide	$C_{14}H_{14}N_2O_6S$	338.33576

Further, we calculated the binding affinities of the 15 ligands (including nimesulide) to the surface of COX-1 and COX-2 enzymes. Nimesulide is a relatively COX-2 selective ligand, but linking of the other ligands to COX-1 enzyme may be interesting too.

The structures of enzymes were retrieved from Protein Data Bank[30,31,32] in PDB format with Chem 3D's "Online Find Structure from PDB id" option. The water molecules, other small molecules, like solvent molecules and other relics of the isolation and crystallization procedures were removed. X-ray crystallography usually does not locate hydrogens, hence most PDB files do not include them. But hydrogens, particularly those that can form hydrogen bonds are important in binding ligands, so hydrogens were added to backbone N, and to amine and hydroxyl side chains. Atoms were renumbered, and PDBQT files generated with AutoDock Tools 1.5.6.

Tables 9 and *10* present the calculated binding affinities in descending order for the ligands and the two enzymes. Rmsd/ub and rmsd/lb values are also shown.

We used the following energy minimization parameters[11,12] : Conjugate Gradients optimization algorithm, 200 total number of steps, stop if energy difference is less than 0.1 kcal/mol.

Table 9. The calculated binding affinities greater than for nimesulide, in descending order for Cyclooxygenase-1 (COX-1).

Enzyme-Ligand (ligand's energies with Babel, kcal/mol)	Binding Affinity [kcal/mol]	rmsd/ub [Å]	rmsd/lb [Å]
COX1_71364034_uff_E=729.52	-8.9	0	0
COX1_71364034_uff_E=729.52	-8.9	5.627	3.618
COX1_10020363_uff_E=717.74	-8.8	0	0
COX1_71364034_uff_E=729.52	-8.8	6.54	4.017
COX1_71771973_uff_E=1249.52	-8.6	0	0
COX1_9884264_uff_E=729.21	-8.6	0	0
COX1_71364034_uff_E=729.52	-8.5	9.832	7.52
COX1_71364034_uff_E=729.52	-8.5	11.877	9.105
COX1_71364034_uff_E=729.52	-8.5	5.479	3.512
COX1_71771973_uff_E=1249.52	-8.5	26.766	24.904
COX1_71771973_uff_E=1249.52	-8.5	26.905	25.149
COX1_10020363_uff_E=717.74	-8.4	49.274	47.526
COX1_40489928_uff_E=686.56	-8.4	0	0
COX1_40489928_uff_E=686.56	-8.4	6.153	2.924
COX1_71365841_uff_E=721.11	-8.4	0	0
COX1_71365841_uff_E=721.11	-8.4	9.387	6.84
COX1_10020363_uff_E=717.74	-8.3	49.062	47.206
COX1_10020363_uff_E=717.74	-8.3	3.327	2.66
COX1_40489928_uff_E=686.56	-8.3	11.074	8.37
COX1_71364034_uff_E=729.52	-8.3	13.191	10.23
COX1_71364034_uff_E=729.52	-8.3	11.028	8.725
COX1_71365841_uff_E=721.11	-8.3	8.874	7.141
COX1_9884264_uff_E=729.21	-8.3	48.916	46.889
COX1_10020363_uff_E=717.74	-8.2	6.471	3.237
COX1_10020363_uff_E=717.74	-8.2	47.894	46.133
COX1_71364034_uff_E=729.52	-8.2	9.6	5.341
COX1_71365841_uff_E=721.11	-8.2	8.869	6.743
COX1_71771973_uff_E=1249.52	-8.2	17.235	15.61
COX1_9884264_uff_E=729.21	-8.2	49.667	47.225
COX1_10902415_uff_E=719.40	-8.1	0	0
COX1_11822507_uff_E=715.58	-8.1	0	0

COX1_9884264_uff_E=729.21	-8.1	19.923	17.616
COX1_9884264_uff_E=729.21	-8.1	4.643	3.383
COX1_9884264_uff_E=729.21	-8.1	48.342	46.788
COX1_10020363_uff_E=717.74	-8	4.558	3.239
COX1_10902415_uff_E=719.40	-8	5.664	2.509
COX1_11822507_uff_E=715.58	-8	48.441	46.521
COX1_4495_uff_E=726.56	-8	0	0

Table 10. The calculated binding affinities greater than for nimesulide, in descending order for Cyclooxygenase-2 (COX-2).

Enzyme-Ligand (ligand's energies with Babel, kcal/mol)	Binding Affinity [kcal/mol]	rmsd/ub [Å]	rmsd/lb [Å]
COX2_10020363_uff_E=717.74	-8.9	6.735	4.247
COX2_10020363_uff_E=717.74	-8.9	0	0
COX2_10020363_uff_E=717.74	-8.8	27.241	24.927
COX2_10020363_uff_E=717.74	-8.8	34.578	31.98
COX2_9884264_uff_E=729.21	-8.6	0	0
COX2_71364034_uff_E=729.52	-8.6	6.05	3.247
COX2_71364034_uff_E=729.52	-8.6	34.496	33.242
COX2_71364034_uff_E=729.52	-8.6	0	0
COX2_71364034_uff_E=729.52	-8.5	34.864	33.455
COX2_71364034_uff_E=729.52	-8.4	20.365	17.335
COX2_10020363_uff_E=717.74	-8.4	34.445	31.683
COX2_71364034_uff_E=729.52	-8.3	5.588	2.806
COX2_71364034_uff_E=729.52	-8.3	34.965	33.105
COX2_71364034_uff_E=729.52	-8.3	18.933	16.92
COX2_69899876_uff_E=742.87	-8.3	0	0
COX2_4495_uff_E=726.56	-8.3	0	0

Using the 2Å cutoff value, the two metrics used for RMSD (summarized in *Table 9* and *10*) indicate that 7 predictions for tested compounds are very accurate in case of COX-1 (71364034, 71771973, 9884264, 40489928, 71365841, 10902415, 11822507), and 4 predictions (10020363, 9884264, 71364034, 69899876) are very accurate for COX-2. Therefore, results indicate that 9 compounds are better ligands of COX-1 and COX-2 than nimesulide (4495) because they require less energy for binding.

We used the default docking parameters: 9 number of binding modes, and exhaustiveness (thoroughness of search): 8. Almost all binding scenarios to COX-2 of 10020363 (N-[2-(2,4-difluorophenoxy)-4-nitrophenyl]methanesulfonamide) and 71364034 (1,1,1-trifluoro-N-(4-nitro-2-phenoxyphenyl)methanesulfonamide) present higher binding affinity, than nimesulide. This suggests that these substances will successfully substitute nimesulide:

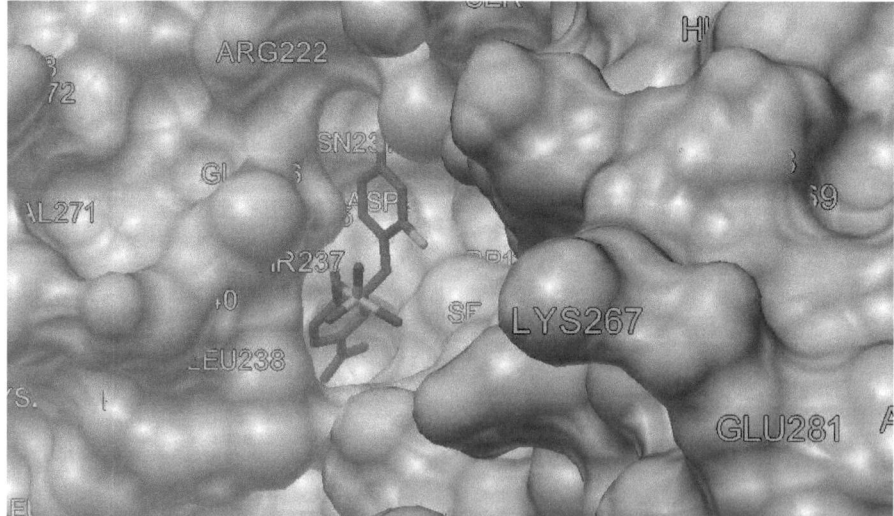

10020363   71364034

Fig. 13. An instance of binding of CID 10020363 to the COX-2's surface

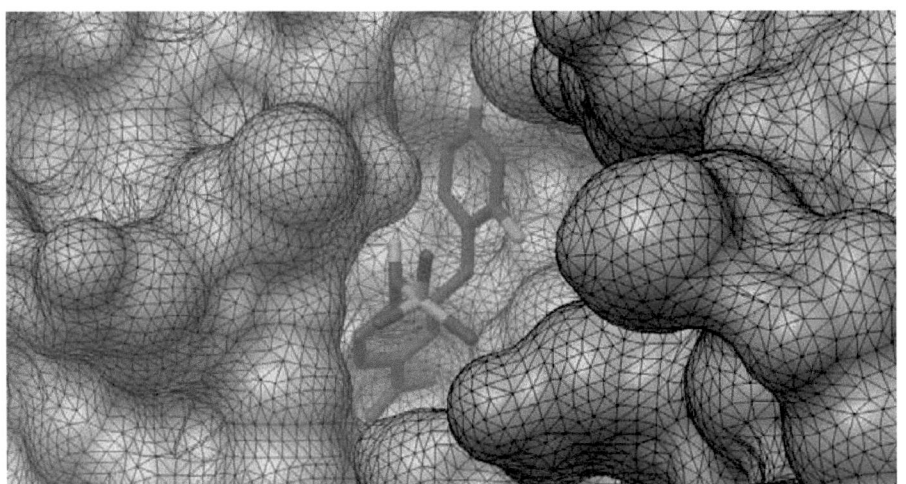

Fig. 14. The surface of the COX-2 (Gouraud shades polygons, AutoDock)

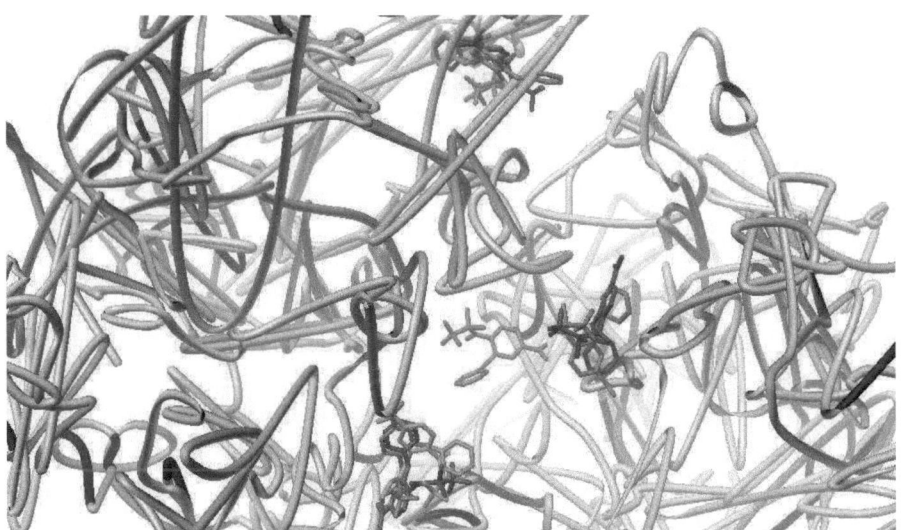

Fig. 15. Nine binding scenarios of CID 71364034 to COX-2 (represented with beaded ribbons)

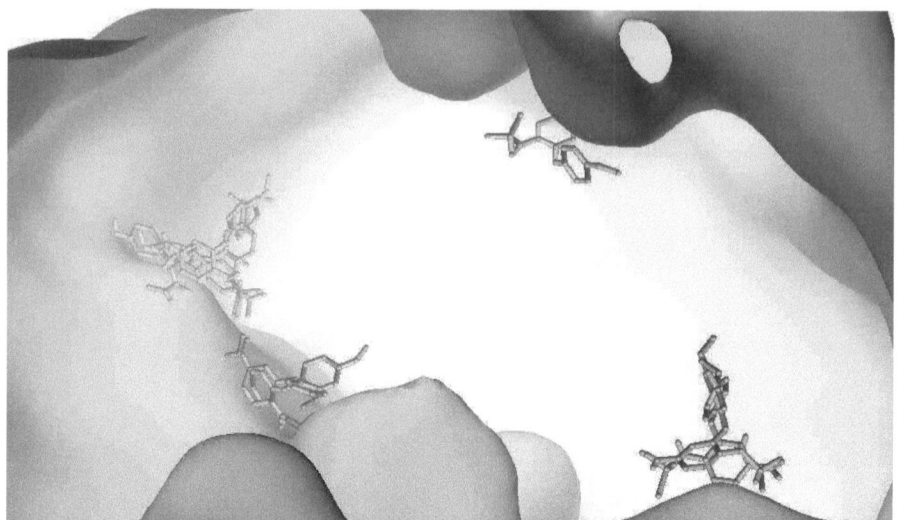

Fig. 16. Coarse molecular surface of the enzyme COX-2 with 9 CID 9884264 molecules

Substance with CID 9884264 N-[2-(4-methoxyphenoxy)-4-nitrophenyl]methane-sulfonamide does not contain fluorine, but a methoxy group. Substance 71771973 (N-methylsulfonyl-N-(4-nitro-2-phenoxy-phenyl)methane-sulfonamide) binds well with COX-1:

9884264   71771973

Other possible "candidates" are:

40489928

71365841

69899876

11822507

10902415

In conclusion, compounds with PubChem ID: 71364034, 71771973, 9884264, 40489928, 71365841, 10902415, 11822507, 10020363 and 69899876 have better binding affinity to cyclooxygenase enzymes (COX-1 and COX-2) than nimesulide, they present the correct bound structure prediction, so they seem to be good substitutes for nimesulide. Further investigations are needed to establish their pharmacodynamic properties and toxicity.

5. Paracetamol, another controversial NSAID

Paracetamol (acetaminophen, PubChem Compound ID 1983) is a worldwide-used medicine for its analgesic and antipyretic actions. It has a spectrum of action similar to that of non-steroidal anti-inflammatory drugs (NSAIDs) and resembles particularly the cyclooxygenase-2 (COX-2) selective inhibitors.

1983

Paracetamol is, on average, a weaker analgesic than NSAIDs or COX-2 selective inhibitors, but is often preferred because of its better tolerance. Despite the similarities to NSAIDs, the paracetamol mode of action has been uncertain, but it is now generally accepted that it inhibits COX-1 and COX-2 through metabolism by the peroxidase function of these isoenzymes[36].

It is commonly used for the relief of headaches and other minor aches and pains and is a major ingredient in numerous cold and flu remedies. In combination with opioid analgesics, paracetamol can also be used in the management of more severe pain such as post-surgical pain and in providing palliative care in advanced cancer patients[37].

In recommended doses and for a limited course of treatment, the side effects of paracetamol are mild to non-existent. While generally safe for use at recommended doses, acute overdoses of paracetamol can potentially cause fatal liver damage.

---

[36] G.G. Graham, M.J. Davies, R.O. Day, A. Mohamudally, K.F. Scott, *Inflammopharmacology*, **2013**, *21* (3):201-232.

[37] Scottish Intercollegiate Guidelines Network (SIGN) (**2008**). "6.1 and 7.1.1". Guideline 106: Control of pain in adults with cancer. Scotland: National Health Service (NHS).

According to the US Food and Drug Administration, *"Acetaminophen can cause serious liver damage if more than directed is used"*[38].

Because of this, other similar compounds must be found to replace this active substance, and to enlarge the spectrum of prostaglandin-endoperoxide synthase inhibitors used today in therapeutics.

Using the Chemical Structure Search of the PubChem Compound Database and a threshold ≥ than 95% for the similar structures criteria, we detected 37 compounds that meet these criteria. We have calculated the binding affinities for the ligands (including paracetamol) to the surface of COX-1 and COX-2 enzymes.

32 substances (CID = 21039831, 54373747, 28739598, 58156, 6435661, 12541316, 70185934, 13565130, 62679285, 190616, 837933, 21304453, 56990904, 62679286, 64794047, 70503032, 72907, 836029, 64795168, 29520, 836551, 10979586, 23424464, 28739570, 64780675, 64793306, 1725710, 64781061, 4301564, 57880597, 66874, 64782388) are better ligands for COX-1 than paracetamol, 28 compounds (CID = 21039831, 54373747, 58156, 64780675, 70185934, 23424464, 72907, 21304453, 836029, 6435661, 10979586, 12541316, 13565130, 70503032, 29520, 28739598, 62679286, 64781061, 1725710, 3014069, 21931829, 66874, 190616, 62679285, 64782388, 74325, 28739570, 70458847) are better ligand for COX-2. The best 7 ligands are shown in *Table 11*, together with paracetamol (CID 1983).

Table 11. The best 7 ligands with a similarity threshold ≥ 95% and the paracetamol (CID 1983)

PubChem CID	IUPAC Name	Molecular Formula	MW [g/mol]
21039831	N-(4-hydroxyphenyl)octadeca-2,4,6,8,10,12,14,16-octaynamide	$C_{24}H_9NO_2$	343.33376
54373747	N'-(4-hydroxyphenyl)but-2-enediamide	$C_{10}H_{10}N_2O_3$	206.198
28739598	(2E,4E)-N-(4-hydroxyphenyl)hexa-2,4-dienamide	$C_{12}H_{13}NO_2$	203.23712
58156	N-(4-hydroxyphenyl)-3-methylbut-2-enamide	$C_{11}H_{13}NO_2$	191.22642

---

[38] US FDA. Page Last Updated: April 28, 2014. Acetaminophen Information Page. Accessed May 5, **2014**.

64780675	N-(3-fluoro-4-hydroxyphenyl)but-3-enamide	$C_{10}H_{10}FNO_2$	195.190303
70185934	(Z)-N'-(4-hydroxyphenyl)but-2-enediamide	$C_{10}H_{10}N_2O_3$	206.198
23424464	(E)-N-(4-hydroxyphenyl)but-2-enamide	$C_{10}H_{11}NO_2$	177.19984
1983 (Paracetamol)	N-(4-hydroxyphenyl)acetamide	$C_8H_9NO_2$	151.16256

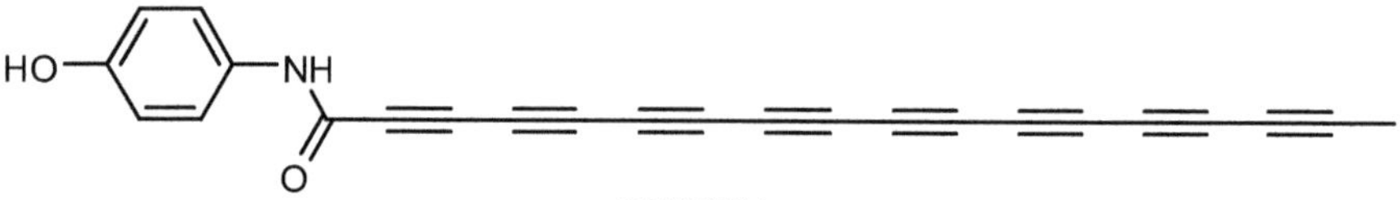

21039831

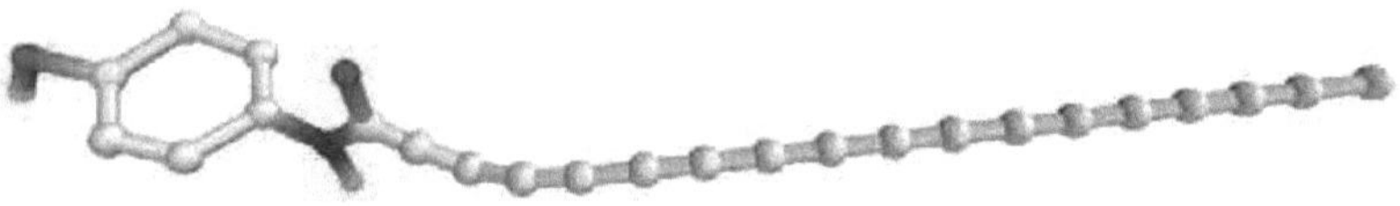

Fig. 17. The N-(4-hydroxyphenyl)octadeca-2,4,6,8,10,12,14,16-octaynamide, CID 21039831 (Open Babel UFF molecular energy minimization)

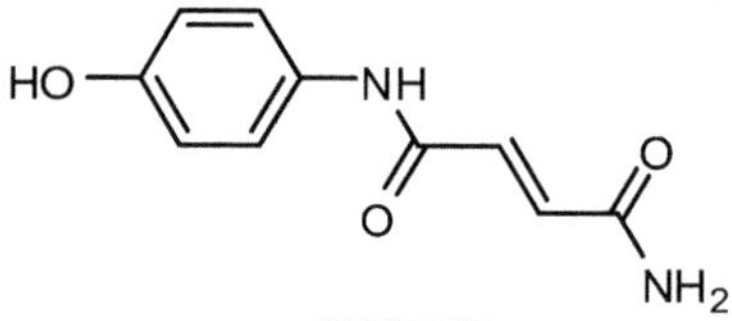

54373747

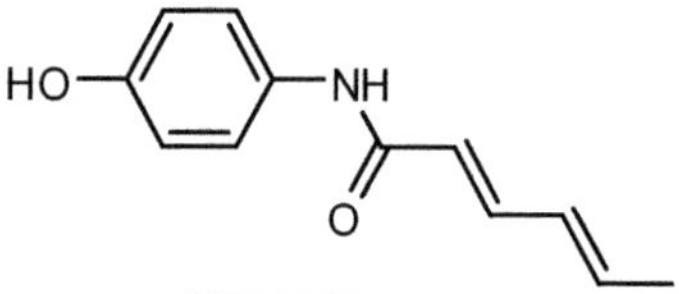

28739598

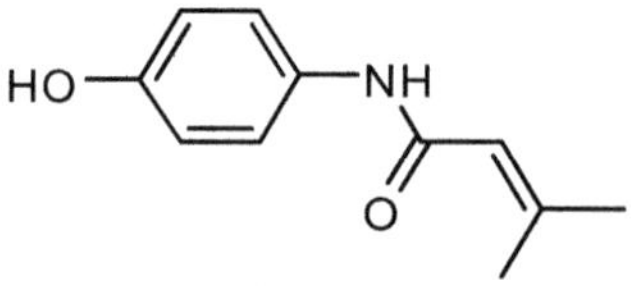

58156

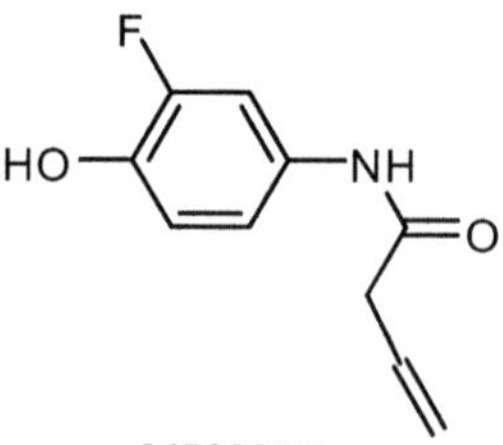

64780675

70185934

23424464

Table 12 and *Table 13* present the calculated binding affinities in descending order for the ligands and the two enzymes (COX-1 and COX-2), only for rmsd/ub and rmsd/lb = 0.

The structures of enzymes were retrieved from Protein Data Bank[30,31,32] in PDB format with Chem 3D's "Online Find Structure from PDB id" option. The water molecules, other small molecules, like solvent molecules and other relics of the isolation and crystallization procedures were removed.

X-ray crystallography usually does not locate hydrogens, hence most PDB files do not include them. But hydrogens, particularly those that can form hydrogen bonds, are important in binding ligands, so hydrogens were added to backbone N, and to amine and hydroxyl side chains. Atoms were renumbered, and PDBQT files generated with AutoDock Tools 1.5.6.

We used the following energy minimization parameters: Conjugate Gradients optimization algorithm, 200 total number of steps, stop if energy difference is less than 0.1 kcal/mol.

The virtual screening results show that 32 compounds are strong inhibitors of COX-1, and 28 compounds for COX-2. (*Table 12* and *13*.)

Table 12. The calculated binding affinities greater than for paracetamol, in descending order for Cyclooxygenase-1 (COX-1).

Enzyme-Ligand (ligand's energies with Babel, kcal/mol)	Binding Affinity [kcal/mol]	rmsd/ub [Å]	rmsd/lb [Å]
COX1_21039831_uff_E=7278.98	-8.3	0	0
COX1_54373747_uff_E=135.65	-7.4	0	0
COX1_28739598_uff_E=434.80	-7.1	0	0
COX1_58156_uff_E=155.62	-6.9	0	0
COX1_6435661_uff_E=148.61	-6.8	0	0

43

COX1_12541316_uff_E=136.97	-6.8	0	0
COX1_70185934_uff_E=182.81	-6.8	0	0
COX1_13565130_uff_E=176.63	-6.7	0	0
COX1_62679285_uff_E=469.54	-6.7	0	0
COX1_190616_uff_E=420.09	-6.6	0	0
COX1_837933_uff_E=171.92	-6.6	0	0
COX1_21304453_uff_E=171.92	-6.6	0	0
COX1_56990904_uff_E=469.84	-6.6	0	0
COX1_62679286_uff_E=128.27	-6.6	0	0
COX1_64794047_uff_E=292.45	-6.6	0	0
COX1_70503032_uff_E=211.21	-6.6	0	0
COX1_72907_uff_E=774.64	-6.5	0	0
COX1_836029_uff_E=197.19	-6.5	0	0
COX1_64795168_uff_E=125.31	-6.5	0	0
COX1_29520_uff_E=465.06	-6.4	0	0
COX1_836551_uff_E=145.19	-6.4	0	0
COX1_10979586_uff_E=469.84	-6.4	0	0
COX1_23424464_uff_E=121.50	-6.4	0	0
COX1_28739570_uff_E=151.06	-6.4	0	0
COX1_64780675_uff_E=471.47	-6.4	0	0
COX1_64793306_uff_E=130.47	-6.4	0	0
COX1_1725710_uff_E=132.49	-6.3	0	0
COX1_64781061_uff_E=122.21	-6.3	0	0
COX1_4301564_uff_E=413.18	-6.2	0	0
COX1_57880597_uff_E=111.05	-6.2	0	0
COX1_66874_uff_E=122.15	-6.1	0	0
COX1_64782388_uff_E=123.73	-6.1	0	0
COX1_1983_uff_E=414.99	-6	0	0

Table 13. The calculated binding affinities greater than for paracetamol, in descending order for Cyclooxygenase-2 (COX-2).

Enzyme-Ligand (ligand's energies with Babel, kcal/mol)	**Binding Affinity** [kcal/mol]	**rmsd/ub** [Å]	**rmsd/lb** [Å]
COX2_21039831_uff_E=7278.98	-8	0	0
COX2_54373747_uff_E=135.65	-7.4	0	0

COX2_58156_uff_E=155.62	-7.3	0	0
COX2_64780675_uff_E=471.47	-7.2	0	0
COX2_70185934_uff_E=182.81	-7.2	0	0
COX2_23424464_uff_E=121.50	-7.1	0	0
COX2_72907_uff_E=774.64	-7	0	0
COX2_21304453_uff_E=171.92	-7	0	0
COX2_836029_uff_E=197.19	-6.9	0	0
COX2_6435661_uff_E=148.61	-6.9	0	0
COX2_10979586_uff_E=469.84	-6.9	0	0
COX2_12541316_uff_E=136.97	-6.9	0	0
COX2_13565130_uff_E=176.63	-6.9	0	0
COX2_70503032_uff_E=211.21	-6.9	0	0
COX2_29520_uff_E=465.06	-6.8	0	0
COX2_28739598_uff_E=434.80	-6.8	0	0
COX2_62679286_uff_E=128.27	-6.8	0	0
COX2_64781061_uff_E=122.21	-6.8	0	0
COX2_1725710_uff_E=132.49	-6.7	0	0
COX2_3014069_uff_E=417.03	-6.7	0	0
COX2_21931829_uff_E=462.37	-6.7	0	0
COX2_66874_uff_E=122.15	-6.6	0	0
COX2_190616_uff_E=420.09	-6.6	0	0
COX2_62679285_uff_E=469.54	-6.6	0	0
COX2_64782388_uff_E=123.73	-6.6	0	0
COX2_74325_uff_E=120.63	-6.5	0	0
COX2_28739570_uff_E=151.06	-6.5	0	0
COX2_70458847_uff_E=425.01	-6.5	0	0
COX2_1983_uff_E=414.99	-6.4	0	0

The RMSD cutoff of 2Å is usually used as criteria of the correct bound structure prediction[24]. Using the same cutoff value, the two metrics used for RMSD (summarized in *Table 12* and *13*) indicate that all predictions for tested compounds are very accurate, but in case of COX-1 three predictions (CID 21039831, 54373747, 28739598), and in case of COX-2 six predictions (CID 21039831, 54373747, 58156, 64780675, 70185934, 23424464) have a binding affinity greater than 7 kcal/mol.

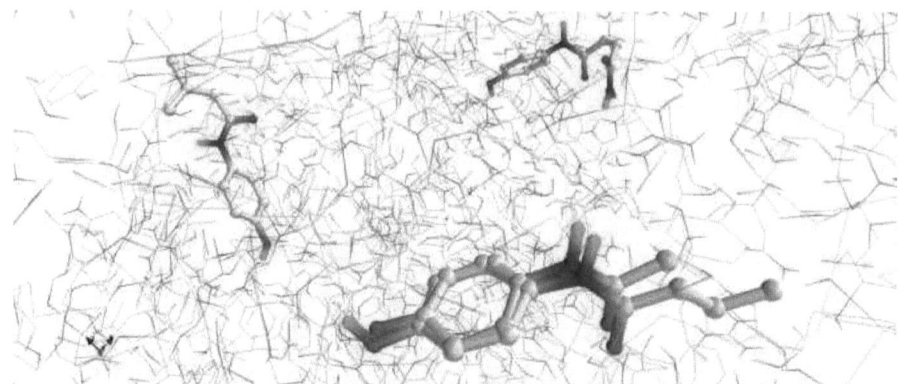

Fig. 18. Molecular docking and position of paracetamol and 3 "top" ligands in protein target (COX-2)

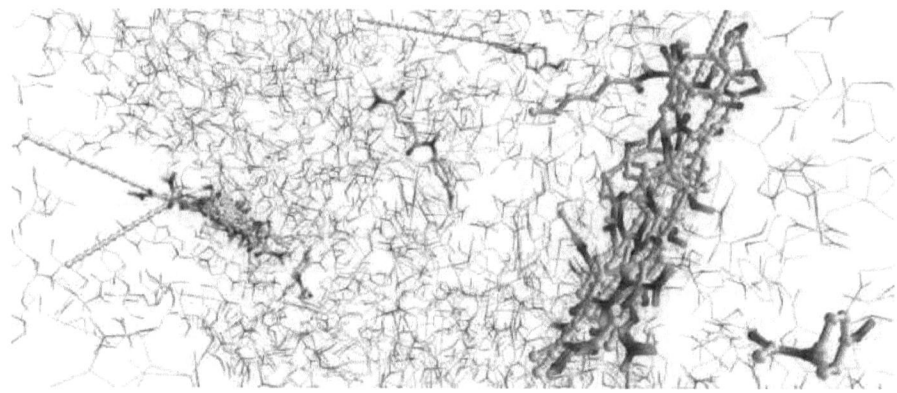

Fig. 19. All (36) binding scenarios of CID 21039831, 54373747, 28739598, paracetamol and COX-1

Therefore, results indicate that 7 compounds with binding affinity greater than 7 kcal/mol, are much better ligands of COX-1 and COX-2 than paracetamol (CID 1983). They have much better binding affinity to cyclooxygenase enzymes (especially to COX-2) than paracetamol, and they present the correct bound structure prediction.

This suggests that these substances will successfully substitute paracetamol. Further investigations are needed to establish their pharmacodynamic properties and toxicity.

We used the default docking parameters: 9 number of binding modes, and exhaustiveness (thoroughness of search): 8. Larger values increase the probability of finding the global minimum, but also extend the computational time. Increasing the exhaustiveness value increases the time linearly and decreases the probability of not finding the minimum exponentially.

Fig. 20. Chain A and chain B in COX-1

Other 22 very good ligands for COX-2 are:

72907	21304453	836029

47

6435661

10979586

12541316

13565130

70503032

29520

28739598

64781061

62679286

1725710

3014069

21931829

48

66874

190616

62679285

64782388

74325

28739570

70458847

6. Brodifacoum and other "super-warfarines"

Brodifacoum (PubChem Compound Identifier, CID=54680676) is a coumarin derived rodenticide. It was first introduced in 1975 to deal with the public health problem of warfarin resistant rodents.

Brodifacoum is a weak acid which does not readily form water soluble salts. It does not lose activity after 30 days in direct sunlight. This rodenticide is effective against warfarin resistant rats. Currently, it is registered for the control of rats and mice in and around farm structures, households, and domestic dwellings, inside transport vehicles, commercial transportation facilities, industrial areas, sewage systems, aircraft, ships, boats, railway cars, and food processing, handling and storage areas. Products containing brodifacoum are available to the general public and application may be made as often as necessary. Brodifacoum is formulated as meal bait, paraffinized pellets, rat and mouse bait ready-to-use place packs, and paraffin blocks. All end-use products contain 0.005 percent active ingredient[39].

54680676

Brodifacoum was made a "restricted use" pesticide in 2008 by EPA, meaning it can only be used by certified pesticide applicators. The product remains on the market for public use.

---

[39] American Bird Conservancy, http://www.abcbirds.org/abcprograms/policy/toxins/Profiles/brodifacoum.html. **Retrieved 9 April 2015.**

Brodifacoum is absorbed through the gut and works by preventing the normal clotting of blood, leading to fatal hemorrhage. It inhibits coagulation by antagonizing the action of vitamin K. Warfarin prevents the recycling of vitamin K by blocking vitamin K epoxide reductase (VKOR) activity, thus creating a functional vitamin K deficiency. Inadequate gamma-carboxylation of vitamin K-dependent coagulation proteins interferes with the coagulation cascade, which inhibits blood clot formation.

Death usually occurs through gastric hemorrhage. It is retained in the tissues at high rates, sometimes remaining in organ systems during the entire lifetime of an exposed animal. In a study that measured the retention of radioactive brodifacoum in the livers of single-dosed rats, 34% of the single dose is found in the liver after 13 weeks, and 11% of the dose remained in the liver for 104 weeks, approaching the normal lifespan of a rat[39].

It is highly effective at small doses - usually a rodent ingests a fatal dose after a single feeding and will die within 4-5 days. The greatest risk to wildlife from brodifacoum is secondary poisoning. Rodents continue to eat poisoned bait, so at the time of death the amount of brodifacoum present in their bodies is many times the amount required to kill them. Non-target wildlife such as predators and scavengers may then consume rodents that have ingested large doses of brodifacoum. It can take as little as one poisoned rodent, or a predator may accumulate enough brodifacoum after consuming several poisoned prey items, to induce life-threatening or fatal effects. A single dose of brodifacoum can depress blood clotting for months in some animals, including birds. Stress or slight wounds incurred in the field, such as small scratches that normally occur when a raptorial bird captures its prey, are often sufficient to cause a fatal hemorrhage.

It is highly toxic to aquatic organisms, mammals and birds. Due to its extremely low solubility and usage patterns however, it is assumed that not enough brodifacoum would dissolve in water to create a hazard to aquatic non-target animals. Products used in sewers are water-resistant paraffinized blocks and are not expected to dissolve in water. But hundreds of avian and other wildlife mortalities have been reported across North America[40].

---

[40] U.S. National Library of Medicine, TOXNET®: Toxicology Data Network, http://toxnet.nlm.nih.gov/cgi-bin/sis/search/a?dbs+hsdb:@term+@DOCNO+3916. Retrieved 9 April 2015

Because of this, it may be interesting to find similar compounds in order to enlarge the spectrum of VKOR inhibitors used today in pest control. We used the Similar Compounds search type of the Chemical Structure Search of the PubChem Compound Database, to locate records that are similar to the chemical structure of brodifacoum, using pre-specified similarity thresholds. Using the threshold $\geq$ than 95% for the similar structure criteria, we found 14 compounds that meet these criteria. In accordance with our calculations and molecular docking simulations, two of these compounds have a better binding affinity to vitamin K epoxide reductase enzyme than brodifacoum, but the binding energy of the other 12 substances is also high, having identical or lower lipophilicity, consequently they will eliminate faster, possibly lacking a part of the adverse effects.

We calculated the binding affinities for the ligands (including for brodifacoum) to the surface of VKOR enzyme. Fourteen substances (with PubChem Compound ID = 54721414, 54721409, 54721416, 54702967, 54702968, 56999900, 54676115, 54716798, 56610083, 56638536, 56629283, 57166454, 57186316, 54736729) are excellent ligands for VKOR. These ligands are shown in *Table 14* and below, together with brodifacoum (CID 54680676).

Table 14. The 14 ligands with a similarity threshold $\geq$ 95%, and brodifacoum (54680676)

PubChem CID	IUPAC Name	Molecular Formula	MW [g/mol]	XlogP3
54721414	3-[3-[4-[[4-(4-bromophenyl)phenyl]methoxy]phenyl]-1,2,3,4-tetrahydronaphthalen-1-yl]-4-hydroxychromen-2-one	$C_{38}H_{29}BrO_4$	629.53846	9.1
54721409	3-[3-[4-[4-[(4-bromophenyl)methoxy]phenyl]phenyl]-1,2,3,4-tetrahydronaphthalen-1-yl]-4-hydroxychromen-2-one	$C_{38}H_{29}BrO_4$	629.53846	9.1
54680676	3-[3-[4-(4-bromophenyl)phenyl]-1,2,3,4-tetrahydronaphthalen-1-yl]-4-hydroxychromen-2-one	$C_{31}H_{23}BrO_3$	523.41652	7.6
54721416	3-[3-[4-[[4-(4-bromophenyl)phenyl]methyl]phenyl]-1,2,3,4-tetrahydronaphthalen-1-yl]-4-hydroxychromen-2-one	$C_{38}H_{29}BrO_3$	613.53906	9.6

54702967	3-[3-[4-[2-(4-bromophenyl)ethyl]phenyl]-1,2,3,4-tetrahydronaphthalen-1-yl]-4-hydroxychromen-2-one	$C_{33}H_{27}BrO_3$ 551.46968	8.2
54702968	3-[3-[4-[(4-bromophenyl)methoxy]phenyl]-1,2,3,4-tetrahydronaphthalen-1-yl]-4-hydroxychromen-2-one	$C_{32}H_{25}BrO_4$ 553.4425	7.5
56999900	3-[[4-(4-bromophenyl)phenyl]-cyclohexylmethyl]-4-hydroxychromen-2-one	$C_{28}H_{25}BrO_3$ 489.4003	7.6
54676115	3-[(1S,3R)-3-[4-(4-bromophenyl)phenyl]-1,2,3,4-tetrahydronaphthalen-1-yl]-4-hydroxychromen-2-one	$C_{31}H_{23}BrO_3$ 523.41652	7.6
54716798	3-[(1R,3R)-3-[4-(4-bromophenyl)phenyl]-1,2,3,4-tetrahydronaphthalen-1-yl]-4-hydroxychromen-2-one	$C_{31}H_{23}BrO_3$ 523.41652	7.6
56610083	3-[(2S,4R)-4-[4-(4-bromophenyl)phenyl]-1,2,3,4-tetrahydronaphthalen-2-yl]-4-hydroxychromen-2-one	$C_{31}H_{23}BrO_3$ 523.41652	7.6
56638536	3-[(2R)-4-[4-(4-bromophenyl)phenyl]-1,2,3,4-tetrahydronaphthalen-2-yl]-4-hydroxychromen-2-one	$C_{31}H_{23}BrO_3$ 523.41652	7.6
56629283	3-[(2S)-4-[4-(4-bromophenyl)phenyl]-1,2,3,4-tetrahydronaphthalen-2-yl]-4-hydroxychromen-2-one	$C_{31}H_{23}BrO_3$ 523.41652	7.6
57166454	3-[1-[4-(4-bromophenyl)phenyl]-2-phenylethyl]-4-hydroxychromen-2-one	$C_{29}H_{21}BrO_3$ 497.37924	6.8
57186316	3-[(2S)-1-[4-(4-bromophenyl)phenyl]-1,2,3,4-tetrahydronaphthalen-2-yl]-4-hydroxychromen-2-one	$C_{31}H_{23}BrO_3$ 523.41652	7.6
54736729	3-[(2R,4R)-4-[4-(4-bromophenyl)phenyl]-1,2,3,4-tetrahydronaphthalen-2-yl]-4-hydroxychromen-2-one	$C_{31}H_{23}BrO_3$ 523.41652	7.6

54721414

54721409

54721416

54

54702967

54702968

56999900

54676115

54716798

56610083

56638536

56629283

57166454

57186316

54736729

Table 15 presents the calculated binding affinities in descending order for the ligands and the enzyme VKOR (PDB code 3KP9 [41]).

The structure of enzyme was retrieved with Chem 3D's Online Find Structure from PDB ID option. The water molecules, other small molecules, like solvent molecules and other relics of the isolation and crystallization procedures were removed.

Table 15. The calculated binding affinities (including brodifacoum, CID = 54680676) in descending order for the enzyme VKOR (3KP9).

Enzyme-Ligand	Binding affinity [kcal/mol]	rmsd/ub [Å]	rmsd/lb [Å]
3KP9_54721414_mmff94_E=259.03	-11.3	0	0
3KP9_54721409_mmff94_E=203.54	-10.6	0	0
3KP9_54680676_mmff94_E=162.68	-10	0	0
3KP9_54721416_mmff94_E=182.96	-9.8	0	0
3KP9_54702967_mmff94_E=172.60	-9.5	0	0
3KP9_54702968_mmff94_E=173.00	-9.5	0	0
3KP9_54721416_mmff94_E=182.96	-9.5	2.003	1.139
3KP9_54716798_mmff94_E=880.70	-9.3	0	0
3KP9_56999900_mmff94_E=1010.91	-9.3	0	0
3KP9_54676115_mmff94_E=880.70	-9.2	0	0
3KP9_56610083_mmff94_E=954.18	-9	0	0
3KP9_56638536_mmff94_E=956.00	-9	0	0
3KP9_56629283_mmff94_E=881.16	-8.7	0	0
3KP9_57166454_mmff94_E=125.89	-8.7	0	0
3KP9_57186316_mmff94_E=884.05	-8.7	0	0
3KP9_54736729_mmff94_E=903.33	-8.1	0	0

Using the 2Å cutoff value, the two metrics used for RMSD (summarized in *Table 15*) indicate that 2 compounds are better ligands of VKOR than brodifacoum (CID 54680676), because they require less energy for binding. This suggests that these substances will successfully substitute brodifacoum. The binding energy of the other 12 substances is also high, having identical or lower lipophilicity, consequently they will eliminate faster, possibly lacking a part of the adverse effects.

---

[41] W. Li, S. Schulman, R.J. Dutton, D. Boyd, J. Beckwith, T.A. Rapoport, *Nature*, **2010**, *463*: 507-512. http://www.rcsb.org/pdb/explore/explore.do?structureId=3kp9

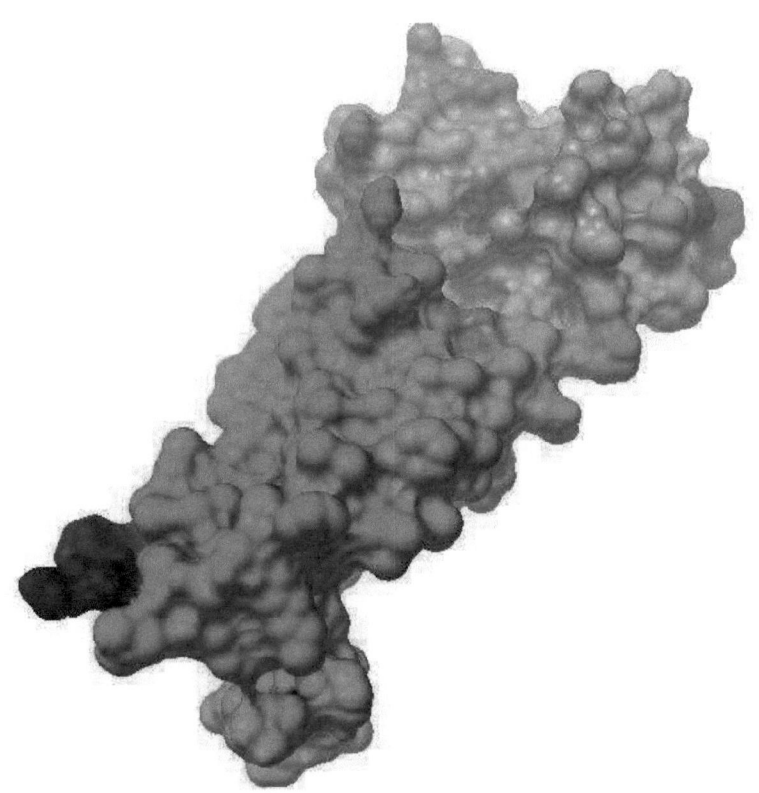

Fig. 21. The enzyme surface (AutoDock)

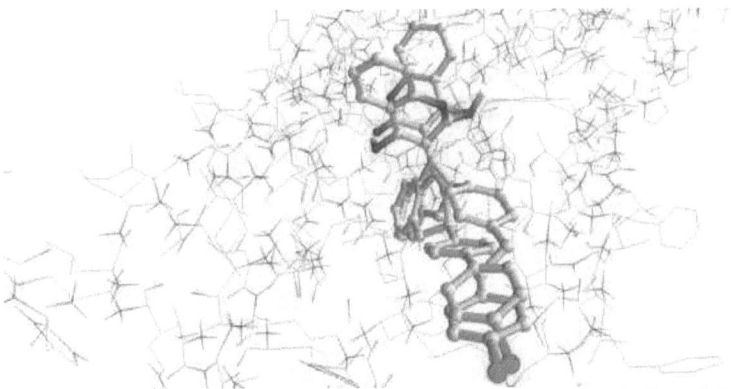

Fig. 22. Molecular docking of brodifacoum (CID 54680676) and the ligand CID 54702968 in protein target (Mayavi).

Compounds with PubChem ID: 54721414, 54721409, 54721416, 54702967, 54702968, 56999900, 54676115, 54716798, 56610083, 56638536, 56629283, 57166454, 57186316 and 54736729 have good binding affinity to vitamin K epoxide reductase enzyme, they present the correct bound structure prediction, so they seem to act as "super-warfarines", and to be good substitutes for brodifacoum. Further investigations are needed to establish their pharmacodynamic properties and toxicity.

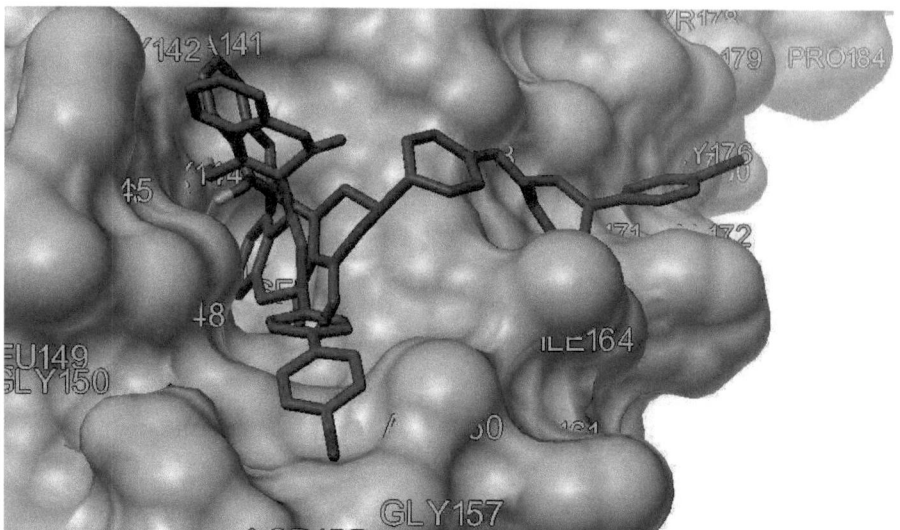

Fig. 23. Ligand CID 54721414 and brodifacoum (CID 54680676) on the binding site of VKOR (Autodock).

Index

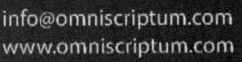

Printed by Books on Demand GmbH, Norderstedt / Germany